Hazard Potential of Chemical Residuals in Food

Hazard Potential of Chemical Residuals in Food

Dr. Ghazala Yaqub and Miss. Haleema Sadia

Delve
PUBLISHING
www.delvepublishing.com

Hazard Potential of Chemical Residuals in Food

Dr. Ghazala Yaqub and Miss. Haleema Sadia

Delve Publishing
2010 Winston Park Drive,
2nd Floor
Oakville, ON L6H 5R7
Canada
www.delvepublishing.com
Tel: 001-289-291-7705
001-905-616-2116
Fax: 001-289-291-7601
Email: orders@arclereducation.com

ISBN: 978-1-77361-102-0 (Hardcover)

ABOUT THE AUTHORS

Dr. Ghazala Yaqub

Dr. Ghazala Yaqub is an Assistant Professor of Environmental Sciences and a member of Academic Council at Kinnaird College for Women, Lahore, a well reputed and prestigious institute in Pakistan. She graduated with a Ph.D. degree in Chemistry in 2011. She did her Post-Doctorate research at the University of Wollongong Australia. She is an author of various research articles and books. Dr. Yaqub was awarded Australian Excellence Award in recognition of her research outputs and community services in 2016. She is recently appointed as an Australian Award Ambassador for Pakistan

Miss. Haleema Sadia

Ms. Haleem Sadia was born and raised in a small village of District Gujrat, Pakistan. She earned her BS (Hons.) degree in Chemistry from the University of Gujrat. She is an independent social worker and a debater and a voice of her hometown and neighborhood countryside. She is a formal motivational speaker for a campaign "The girls should go to the colleges and role of education in the personal development and community progress".

TABLE OF CONTENTS

GLOSSARY

Additives: Non-nutritive material added deliberately to foodstuff, normally in minute amounts to enhance flavor, appearance, storage otherwise texture characteristics. Quantities used in foodstuff are generally synchronized by means of regulations.

Adulteration: The method as a result of which nature or else the quality of a particular material is decreased by:

(i) The addition of a substandard otherwise a foreign material.

(ii) The elimination of crucial constituent. If the physiological utilities of the customer are exaggerated, adulteration of foodstuff can jeopardize fitness.

Alloy: A metallic material composed of two or more elements, as either a compound or a solution. Ordinarily, the components of alloys are metals, although carbon, a nonmetal, is an essential constituent of steel.

Antibiotics: A substance formed as a result of existing organisms so as to inhibit the development of additional organisms, occasionally used like a foodstuff additive otherwise development promoter in several states.

Blanching: Dipping of vegetables otherwise fruits into the sweltering water or else revealing these to vapors, used for a few minutes to exterminate biological as well as an enzymatic activity before processing or freezing.

Brass: An alloy of copper and zinc, of historical and enduring importance because of its hardness and workability. The earliest brass, called calamine brass, was probably made by reduction of mixtures of zinc ores and copper ores.

Bronze: An alloy traditionally composed of copper and tin. Bronze is of exceptional historical interest and still finds wide applications. The proportions of copper and tin vary widely (e.g. from 67 to 95% for copper).
Some modern bronze alloys contain no tin at all, substituting with other metals such as aluminum, manganese and even zinc.

Cast Iron: An alloy of iron that contains 2 to 4% carbon, along with varying amounts of silicon, manganese, and traces of impurities such as sulfur and phosphorus. It is made by reducing iron ore in a blast furnace.

Chemical Addition: Substances additional to unpreserved foodstuff to avoid spoilage by means of prevented expansion of microbes.

Coliform Microorganisms: Collection of aerobic microbes normally presents in feces; their existence specifies shortage of hygiene like in inadequately treated water meant for consumption.

Contaminant: Any material unintentionally added in the direction of foodstuff, which may found in such foodstuff the same as a consequence of the manufacture counting processes performed in animal husbandry, crop husbandry as well as veterinary drugs, processing, preparation, manufacture, treatment, packaging, packing, transport otherwise holding of the provisions or else like a consequence of ecological contagion.

Drinkable Water: Water that is liberated as of pathogenic microbes, in addition, to being pleasant.

Fermentation: Without the utilization of oxygen, breakdown of composites, for example, breakage of sugar by means of the mold to form carbon dioxide as well as alcohols.

Food Poisoning: A broad term used to all intestinal as well as stomach disruptions because of contaminated food by means of some microbes otherwise to their poison.

Fortification: Addition of one otherwise additional nutrient to a foodstuff to make it richer than the natural provisions, for example, Vitamin C addition to fruit drinks.

German Silver: Now known as nickel silver. It contains 10–30% nickel mixed with copper and zinc. This alloy is a typical base for silver-plated ware.

Maintenance Time: The moment subsequent to the removal of a medicine as of the remedy of the creature prior to the animal otherwise its foodstuffs are able to be consumed in favor of individual foodstuff.

Pasteurization: The procedure to kill pathogenic organisms in a foodstuff product by means of warmth on prescribed warmth beneath 1000C generally useful for milk.

Pathogenic: Able to cause illness.

Smoke Point: The temperature on which the breakdown possessions of fat develop into observable like smolder.

Stainless Steel: Stainless steels are a family of over 200 iron-carbon-chromium alloys usually containing 10 to 30% chromium. Stainless steels are also known as inox steel or inox from French "inexorable". In conjunction with low carbon contents, chromium imparts remarkable resistance to corrosion and heat. The different types of stainless steels have different technical properties.

Surface Coating: Surface coatings are the thin films that are functional and often decorative. Surface coatings include paints, drying oils and varnishes, synthetic clear coatings and other products whose primary function is to protect the surface of an object from the near environment.

LIST OF FIGURES

LIST OF TABLES

PREFACE

Life is a practical version of Hazard Potential of Chemical Residuals in Food. It provides the basic knowledge in the field of science, no matter where you are standing- in an office, in a bedroom, in an office, or in a field the knowledge about the food you eat is most important. You are observing the manifestation of Food exposed to other materials in one way or in the other way.

The subject of hazard potential of chemicals has developed and studied by different points of view in the recent past. It is a rigorous and demanding subject. It is informative for all the individuals, irrespective of their field of study, consideration, perspective about food. In this Textbook, I have tried to introduce hazard potential of materials exposed to food, in such a way as to stimulate an interest and inculcate an understanding of the subject. This book has been written with a mechanistic approach to making the subject matter more clear, interesting and understandable. It covers the subject matter ranging from basics through the complete illustration about the mean theme of life.

I have organized the subject matter of this Textbook in a chronological order to best suit your requirements. Each new chapter builds on the foundation laid in the preceding chapters so that when you start with any chapter you find yourself already prepared for the information as well as knowledge. I have tried to cover maximum material in a chapter, yet keeping its length manageable.

Hazard potential of chemical residuals in food is mostly based upon the chemicals exposed to food directly and is, therefore, a pre-requisite to the understanding of hazard potential of chemical residuals in food. An understanding of hazards, source of hazards, effects of hazards and ultimate risks to food safety are therefore discussed in the very first chapter of this book followed by the migration of chemicals exposed to food (Chapter 2) that are very helpful in explaining the migration of chemicals by different food packaging materials to the food. Types of migration of chemicals to food are very important in explaining the migration of different chemicals by different sources. Therefore, a full chapter has been given to the packaging materials (Chapter 7) which is discussed after the exposure of food with artificial chemical residuals and natural chemical residuals because these chemicals provide useful basics and illustrations to explain various packaging materials. Although the chemicals in food or in contact with food are discussed in Chapters 3 and 4, the formal introduction to food industries was postponed till Chapter 8 in which various food industries are discussed. In chapter 5 the evaluation of sensory properties of food has been discussed, and in chapter 6 cleaning and disinfection importance in food processing is explained. The legislations on food contact materials and chemicals in contact with food are discussed in chapter 9 which is followed by one more chapter that is the last chapter of the book Chapter 10 named as prevention from the hazard.

The general description of the chapter and basic knowledge about each chapter is given at the starting of every chapter because it has been found to be an effective way of recognizing the basics about that certain chapter. The ending of each chapter is by the best way to close the chapter and comprehension of the subject attained through study of the chapter. Answers to some questions can be found directly in the text, whereas the other will require some more efforts by reading the text carefully. However, the readers will be able to answer all the questions if they have acquired the comprehension of the subject matter given in the text. Detailed contents, in the beginning, provide the reader with a ready reference to the desired subject matter. Important terms are in bold face and their definition or explanations are given to make the text as clear as possible.

I have used my experience gained by discussions spread over several years to present this job in a simple, systematic, lucid and logical way. I have given an up-to-date account of the subject within the framework of this Textbook. Primarily this book has been written for the graduate students.

You may find some rough spots in the text. Any suggestion to smoothen them will be highly appreciated. You are advised to write down any question that may come to your mind while going through this Textbook. This will help you in understanding the subject and help me in improving the text. I shall also welcome any criticism, comments or suggestions that may contribute toward the improvement of this book.

Author

CHAPTER

1

CHEMICAL RESIDUALS IN FOOD

GENERAL DESCRIPTION

Chemicals may be applied to the foods like pesticides on different vegetables and fruits etc for the benefit to achieve the maximum yield at the minimum cost. These chemicals become a risk, when they are added above a certain level to the food directly or indirectly, leave their impacts to the body, by which these contaminated food are consumed. A variety of chemical residues has different hazardous impacts on food, by which different diseases can occur in humans like nausea, dizziness, dermatitis, vomiting, stomach illness, oesophageal infection and some are dangerous to the extent that these can be carcinogenic. After chronic exposure to the chemicals, the effect of these chemicals can be an immediate or long term associated with their potential. The type of illness, it completely depends upon the possible hazards.

HAZARDS

Any physical, biological or chemical substance in food that may cause adverse health effects, when they are present above a certain level.

Classes of Hazard in Considering to Primary Production

There are three primary types of hazards which are microbiological, chemical and physical. Physical hazards may include; foreign matters such as stones, dust, vegetable matter, needles, and seeds etc. Microbiological hazards include; water and food borne viruses, fungi, bacteria, and parasites etc. Chemical hazards may account environmental contaminants, parasitic residues, veterinary drugs and contaminants from animal feed, natural compounds; such as nitrates, mycotoxins, and nitrites etc. All of these fall under the head of primary classes of hazards. In a threat to the safety of food, all of these play an important role.

CHEMICAL HAZARDS

Chemicals may be added to food at any stage of production, processing, packaging, and eating depends upon the things and properties of the chemicals in contact with these eatables and drinkables. These chemicals may be added naturally or artificially, biologically or chemically, agriculturally or industrially depends upon the food in contact.

Categories of Chemical Hazards

Chemicals hazards can be divided into three categories.

- Intentionally added chemicals.
- Incidentally added chemicals.
- Naturally occurring chemicals.

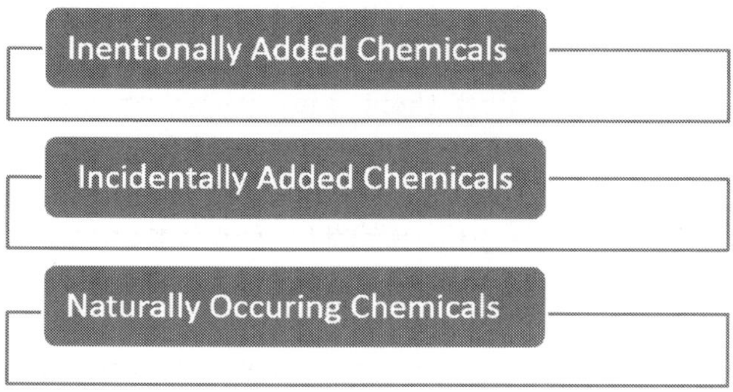

Figure 1.1: Categories of Chemical Hazards

Intentionally Added Chemicals

These chemicals are added during the growth and distribution of the food at some point. These are added according to the certain recommended level. If the amount of these chemicals exceed from a certain level, these chemicals can be very dangerous.

The sources of these chemicals are nitrates, certain preservatives; such as, sulfating agent that can cause allergic reactions in sensitive people. Vitamin A and sodium nitrite are also the examples of these chemicals, which can be toxic in higher concentrations.

Incidentally Added Chemicals

These chemicals become the part of food unintentionally. These might already be in the food ingredients when it is received. Some packaging materials are the indirect source of these chemicals such as sanitizers or inks. Incidental chemicals may include; accidental prohibited substances that may not be allowed at any level such as, poison and insecticides.

The sources of the addition of these chemicals are agricultural chemicals such as pesticides, herbicides that may be very toxic, because of; long exposure of these chemicals to food. Cleaning chemicals, for example, acids and caustics are another sources of incidentally added chemicals. These are also toxic at high concentration. Maintenance chemicals like, lubricants, paints etc are also toxic at higher levels and their contact with food is prohibited.

Naturally Occurring Chemicals

These chemicals are derived from the plants, microorganisms, and animals. These are found naturally in plants prior to or during the harvest. Although many of the chemicals are found in plants naturally these may be harmful or beneficial. The beneficial chemicals are used as medicines and the harmful chemicals may be the toxin in biological origin. These are traditionally categorized as hazard chemicals.

The sources of these chemicals present are certain fish species, molluscan fish, nuts, and corn. These certain fish species, on spoilage, help in the production of toxic levels of histamines and related compounds. Molluscan fish feed on the microscopic organisms that may lead to the production of the toxic compounds such as, domoic acid, that affects people but not affect molluscan fish. Nuts are the source of allergic reactions in sensitive people. Corn also add some toxins to the environment that is, some molds feed on them and produce aflatoxin.

Some biological hazards become the part of the food by food chain and are considered as chemical hazards because of the impacts of the chemicals on the biological system and later on they become the part of the food chain and show their effects chemically such as, veterinary drugs for the treatment of the disease in animals becomes the part of food chain and may show their harmful effects, when exposed to human in the form of food.

SOURCES OF CHEMICAL RESIDUALS IN FOOD

Food provides an ideal environment for the microorganisms because of; enrichment of the suitable environment may be in the form of food, and suitable pH, that is; necessary for the survival of the organism. Meat animals' intestinal tract is full of microorganisms. So, the food products, which are enriched in nutrients, required by microorganisms, and get contaminated. These are generally contaminated primarily by air, dust, sewage; employees, equipment and microorganisms grow on them and release different types of chemicals. These chemicals spoil the food and ultimately become risky for the consumer. Raw food products can also be spoiling by the soil, sewage, live animals, internal meat organs and the external surfaces when they get contaminated by these contaminations.

Sources of Food contamination are multiple, including antibiotics and hormones in meat, natural chemicals in plants and dairy products, as well as microbial contamination that can lead to illness. Foodborne illnesses are very common. Almost 48 million Americans suffer from foodborne diseases each year, and children under age five have the highest incidence of most of these diseases. Microbial contamination of food is regulated and monitored by a number of federal agencies, including, the Department of Agriculture and the Food and Drug Administration. In addition, man-made sources may be the cause of the food contamination by a wide variety of chemicals, typically at low levels. Chemicals in foods may come from the application of pesticides to crops, chemicals in contact with food in any of the stages of food preparation, processing and packaging, and from the transport of industrial chemicals in the environment. A number of environmental contaminants able to accumulate in all types of animals, and are frequently found in meat, fish, poultry, and dairy products. Other chemicals, such as a variety of pesticides and perchlorate, are found in fruits, vegetables, and other agricultural commodities. Some chemicals in food, such as heavy metals and perchlorate, are naturally occurring as well as anthropogenic sources. The actual level of a chemical in the food as well as the amount of the food consumed by individuals determines the health risks from chemicals in food.

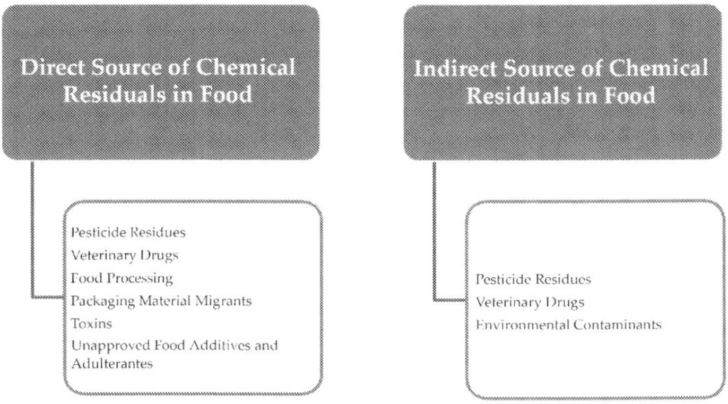

Figure 1.2: Sources of Chemical Residuals in Food

The chemicals become the part of food by many different ways. These chemicals may be added to the environment mainly by direct and indirect sources.

Direct Sources of Chemical Residuals in Food

Direct sources of contamination of food by chemical residuals can be performed by different methods. These are following with their details:

Pesticide Residues

Pesticides, for instance, fungicides, insecticides, or herbicides have become the part of the environment to enhance the yield of the product by applying for these medicines on plants for their better growth and development. So that, by applying these medicines to the plants the microorganisms may kill, that may be harmful to the growth of the plant, and survival also. When these pesticides are sprayed on the plants, no doubt, these kill the organisms such as, harmful fungi, pathogenic bacteria, insects, weeds, pests, and viruses that may cause different illnesses but, in spite of killing these organisms, pesticides also add many new diseases, may be due to improper spraying, or by improper medicines that are sprayed or by inefficiency of the employee etc.

These illnesses may be, because of; the spraying of the pesticides on the food, air, and soil along with the plant or unwanted herbs. If the spray of the pesticides is applied on the soil then it gets contaminated and it also contaminates the water under the soil that after contamination is able to cause diseases in organisms. When these pesticides spray occur in air, these may cause many allergic reactions to the sensitive people and also to the others, because of; exceeding concentrations of pesticides from that certain level, and if the pesticides are applied on the food, then the food gets contaminated may cause different types of diseases in plants and these indirectly added into human and animals and birds respectively by direct and indirect sources, when they are exposed to that food.

For example; if a buffalo eats that plant or food then that contamination will ultimately come in human if they consume that Buffalo, and that's why that food become the part of the food chain and harms many of the organisms that come in contact with them. Directly pesticides harm the plants and food on that plant, and indirectly these may harm the birds, animals, and humans by becoming the part of the food chain.

Veterinary Drugs

Veterinary drugs are used for the service of animals for the treatment of the diseases in animals that they may live and also they can be beneficial for these animals. These have their beneficial effects if these are given in a proper amount and manner, and; if these are not given properly, then these medicines may lead directly to the illness of the animal, and the ultimate effect might be the loss of the animal life.

When the improper medicines are given to the animals or the medicine is given in the improper way, then the internal system of the body of animal may disturb, that leads to the disturbance of the chemical in the body their concentration may change and that leads to the disturbance in pH of the body system that ultimately causes many diseases and these lead the animal to the next level death. If these just affect the animal then these are a direct source of chemical residuals.

Food Processing

During the food processing, like baking, heating, grilling, roasting, hydrolysis or fermentation, and canning certain toxic or undesirable compounds can form.

In smoked and grilled products 'PAHs', in a variety of heat-treated sustenances, specifically coffee and canned or jarred food 'furan', in yeast-fermented alcoholic beverages and other products 'ethyl carbamate', are examples of other processing contaminants.

Cross-contamination may also be caused by food processing like allergens cause non-allergenic food contamination.

Packaging Materials Migrants

The source of chemical contamination by migration of some substances into foods may be, by direct contact of foods with packaging materials. Instances of migrants of health concern may include 4-methyl benzophenone and 2-isopropyl thioxanthone from inks, bisphenol A or phthalates from plastic materials, semicarbazide from a foaming agent or mineral oil from recycled fibers in the plastic gaskets that are used to seal metal lids to glass packaging.

Toxins

Toxins, specifically representing the major concerns in foods, are naturally found substances that are formed by different organisms, with marine bio toxins and mycotoxins. Bacterial toxins 'staphylococcal toxins' or certain plant toxins, like pyrrolizidine alkaloids are instances of food toxins, that can be found in milk, eggs or honey. The bacterial or fungal contamination can be removed by heat treatment but toxins may remain in the food product as contaminants.

Mycotoxins are most frequent toxins that may colonize various crops, are toxic secondary metabolites formed by fungi (molds). Mycotoxins are of concern mainly in cereals, milk, nuts, dried fruit, infant formula, baby food, coffee, wine and fruit juice. Only a few are currently regulated in spite of, the European Union have a more comprehensive list of mycotoxins than many other countries, which includes ochratoxin A, aflatoxins, fumonisins, patulin, zearalenone, deoxynivalenol, and

T-2/HT-2 toxins. In various climates, in different growing and storage conditions, different mycotoxins are frequent.

Marine biotoxins are highly toxic compounds, such as saxitoxin, okadaic acid or ciguatoxin, and domoic acid is produced by phytoplanktons. During harmful algal bloom events, marine biotoxins may cumulate in shellfish or fish, such as scallops or oysters, mussels, clams, to concentrations that may cause serious health risks or even be lethal to humans.

Unapproved Food Additives and Adulterants

Introduction of unapproved food additives to the food or the introduction of wrong food additive by formulation error may cause food adulteration. This results in mislabeled food. Some food adulteration may be harmful to the consumers and some are just to cheat the consumers or mislead them like the addition of high fructose corn syrup to honey. Some unexpected allergic reactions in sensitive individuals may be posed by adulteration of virgin olive oil with hazelnut oil and adulterated chili powders by the use of toxic Sudan dyes.

Indirect Sources of Chemical Residuals in Food

Indirect sources of chemical residuals in food are subsequent:

Pesticide Residues

Recommended dosage of pesticides followed by the Good Agricultural Practices may result in pesticide residues below permissible limits established in a given country. However, the situation gets complicated by the global distribution of food products and global sourcing of raw commodities as the pesticide registrations their permissible limits and uses are different in different countries.

Pesticides play an important role as an indirect source of contamination of food by becoming the part of the food chain from plants to animals and then to humans.

Veterinary Drug Residues

Important reasons behind enforcing their residue limits in foods of animal origin are some specific antibiotics, such as penicillin, that can cause severe allergic reactions in the sensitive individuals. Another important justification for limiting the antibiotic usage in the food-producing animals is to reduce the risk of pathogenic micro-organisms becoming resistant to the antibiotics.

Environmental Contaminants

Environmental contaminants can be anthropogenic or naturally occurring substances present in our air, water or soil. They accumulate in living bodies by entering the food chain. Exposure to some specific contaminants at increased concentration levels may result in critical health conditions. However, the major concern related to the presence of environmental contaminants in foods is their potential endocrine disruption, developmental, carcinogenic and other chronic effects.

The production and the use of polychlorinated biphenyls and other persistent organic pollutants (POPs) have been prohibited for many years, however, they exist in the environment for longer due to their high stability. Polycyclic aromatic hydrocarbons (PAHs) can be found in the environment as a result of industrial pollution or from oil spills. Therefore, oil spill accident in the Gulf of Mexico in 2010, seafood was of great concern due to possible PAHs contamination.

EFFECTS OF CHEMICAL RESIDUALS IN FOOD

A chemical hazard may cause sudden effects or may be associated with potential long-term effects after of chronic exposure to the chemical. One example of an immediate effect is gastrointestinal illness such as nausea, which can be caused by elevated levels of commercial chemicals such as caustic cleaning compounds. Caustic cleaning compounds can also cause burning of the mouth and esophagus.

Ammonia in food contaminated by a refrigerant leak has caused gastrointestinal illness stomachache and nausea and headaches. Sulfites have resulted in diarrhea, headache, difficulty breathing, vomiting, nausea, abdominal pain and cramps in sulfite-sensitive individuals. Examples of long-term effects include impaired cognitive development in children chronically exposed to relatively low levels of lead e.g., in contaminated candy and liver cancer resulting from chronic exposure to the mycotoxin, aflatoxin.

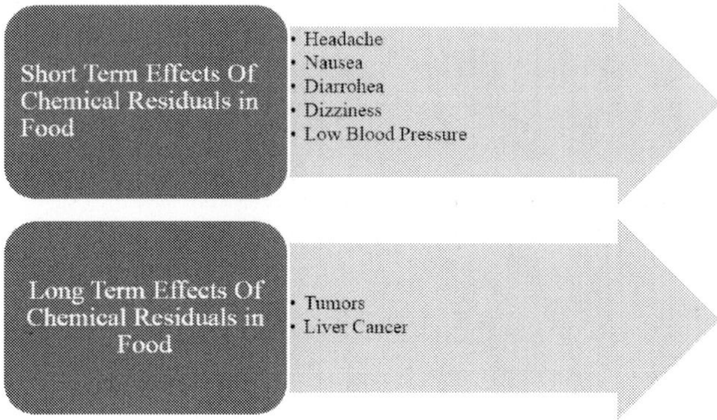

Figure 1.3: Effects of Chemical Residuals in Food

Short-term Effects

A chemical hazard may cause immediate, short-term effects after chronic exposure to the chemical. One example of an immediate effect is gastrointestinal illness such as nausea, which can be caused by increased levels of industrial chemicals (such as caustic cleaning compounds). Caustic cleaning compounds can also cause burning of the mouth and esophagus. Ammonia in food contaminated by a refrigerant leak has caused gastrointestinal illness (stomachache and nausea) and headaches.

Sulfites have resulted in diarrhea, headache, difficulty breathing, vomiting, nausea, abdominal pain and cramps in sulfite-sensitive individuals.

Headache

A headache is described as the twinge start as of the cranium otherwise higher neckline of stiff.

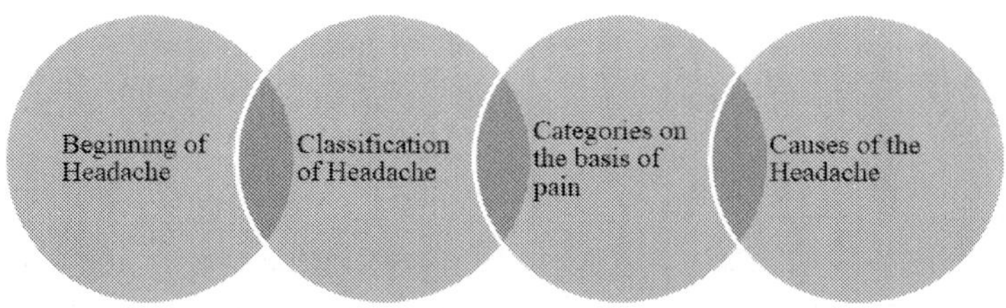

Figure 1.4: A headache

Beginning of Headache

The pain starts as of the structures and tissues that enclose the cranium otherwise the brain as the brain itself has no nerves that offer an increase to the consciousness of pain, pain fibers. The thin film of tissue periosteum that surround muscles, bones so as to cover the sinuses, ears, eyes, and skull, in addition to thin tissues so as to cover the plane of the meninges and brain, nerves, veins, and arteries, all may be irritated otherwise inflamed as well as pose headache. The hurt might be a mild, sharp, throbbing, dull ache, constant, intense, or intermittent.

Classification of Headache

The International Headache Society in 2013 may free, its recent categorization organization in favor of pain. As a result some inhabitants endure as of pain, and because occasionally cure is complicated, it was anticipated so as to the novel arrangement organization would assist health concern experts to construct a further definite analysis and treatment that the kind of pain a tolerant has, and allow improved as well as additional successful options in favor of cure.

The strategies are substantial and the Headache Society suggests so as to health care specialized check with the plans frequently to construct certain of the analysis.

Categories of a headache on the Basis of the Pain

Based upon the source of the pain, there are three major kinds of a headache.

- The main headache
- A minor headache
- Facial pain, cranial neuralgias etc

One class of pain, in addition to so as to supplementary than one kind of pain, might be found at the same time, the submissions also note so as to a patient might have signs that are consonant. All these headaches are related to the chemistry of the brain, as it alters may cause any type of a headache depends upon the change in the chemistry of the brain. These chemicals may disturb the body system by the consumption of food in contact with chemicals, by an individual.

Causes of Headache

- A migraine headache is caused by inflammation or irritation of structures that cover the brain or affects its function. While the brain itself has no pain nerve fibers, everything else from the neck, above the shoulders, skull, and face, can cause a person to have of head pain. Systemic diseases, including infection or dehydration, can have associated headache. These are known as a toxic headache because of the chemicals use more than a certain level that attacks the biological system and causes pain. Changes in circulation and blood flow or trauma can also cause a headache.
- Alternations in brain chemistry may also be related to a headache: medication reactions, drug abuse, and drug withdrawal can all cause pain.
- Every person is different so the history of a headache is important. Recognizing patterns or precipitating (foods were eaten, stress, etc.) factors, in coalition with the physical examination and related symptoms, can help recognize the cause for each individual's specific headache.

Nausea

Nausea is the feeling of a spur to vomit. Nausea can be dangerous and short-lived, or it can be lengthened. When long-lived, it is an exhausting symptom.

Nausea (and vomiting) can be psychological or physical in the start. It can originate from disruptions in the brain or organs of the upper gastrointestinal tract (esophagus, stomach, small intestine, liver, pancreas, and gallbladder). Nausea also may be caused by illnesses of many organs outside of the gastrointestinal system.

Nausea symptoms are mostly strenuous for people to describe. Nausea symptoms are not painful but very intolerable feelings that are felt in the chest, upper abdomen, or back of the throat. Nausea can also be affiliated with: a headache, fever, diarrhea, gas, vomiting, dizziness, lightheadedness, diarrhea, abdominal pain and a general feeling of being sick to one's stomach.

Other causes of nausea comprise side effects from medications, pain, pregnancy and morning sickness, seasickness, viral infection, alcohol toxicity, and disease of the brain, esophagus, stomach, gallbladder, liver, pancreas, and intestine.

Symptoms of Nausea

- Nausea and vomiting are symptoms of an underlying disease and not a specific illness. Nausea is the act of convincing emptying of the stomach. The term "dry heaves" broach to an episode of vomiting where there is no food in the stomach to vomit, and only clear secretions are vomited.
- Vomiting is a violent act in which the stomach nearly turns itself inside out - forcing itself into the lower portion of the esophagus (the tube that relates the mouth to the stomach) during a vomiting episode, expelling food and secretions.

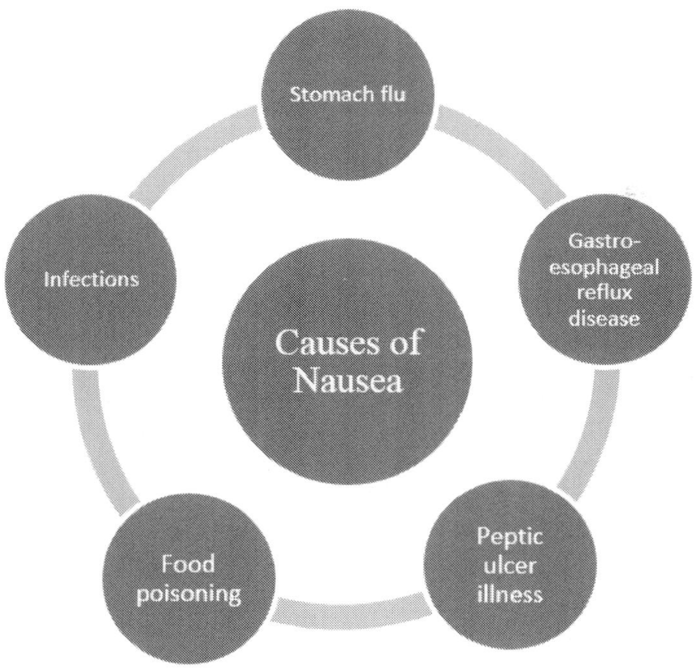

Figure 1.5: Main causes of nausea.

Diarrhea

Diarrhea is an increase in the prevalence of bowel movements or a decrease in the form of stool (greater looseness of stool). Although changes in the frequency of bowel movements and looseness of stools can alter independently of each other, variations often occur in both.

Symptoms of Diarrhea

The symptoms that are correlated with diarrhea depend on the cause and kind of diarrhea.

- If there is a large secretory constituent to diarrhea the bowel movements are frequent and watery. Pain is not common, and there are no signs of swelling.
- Similarly, an osmotic diarrhea is watery, but its main attribute is that once ingestion of food stops (which would include the offending dietary sustenance or substance that is not digested or absorbed) diarrhea stops.

Dizziness

- Vertigo (dizziness) is the sensation of spinning, while lightheadedness is specifically delineated as near fainting, and weakness.
- Some of the conditions that may cause lightheadedness in a patient comprise low blood pressure, high blood pressure, vasovagal syncope, dehydration, heart conditions, medications, diabetes, orthostatic or postural hypotension, endocrine disruptions as well as hyperventilation. Variety in blood pressure may be, because of the disturbances in ion movement by the disruptions of externally consumed food that is in contact with chemicals.
- Vertigo is mostly caused by a problem in the balance centers of the inner ear called the vestibular system and originates the prickle of the room spinning. It may be integrated with vomiting. Signals often are made worse with position changes. Those with specific symptoms and vomiting may need intravenous medication and hospitalization.

- Vertigo is also the presenting symptom in patients with Meniere's disease and acoustic neuroma, conditions that often need a refferal to an ENT specialist. Vertigo may also be a sign of stroke.
- Most often, dizziness or lightheadedness is a temporary condition that resolves spontaneously without a special diagnosis being made.

Introduction to Dizziness (Feeling Dizzy)

The term dizziness is sometimes difficult to understand since it means different things to various people. It is either the sensation of feeling lightheaded as if the individual is frail and will pass out, or it explains vertigo or the sensation of spinning as if the affected person just got off a merry-go-round.

Lightheadedness might be posed by means of disorders of the internal ear as well as the equilibrium midpoints of the mind, while lightheadedness is mostly caused by a decrease in blood supply to the brain. It is significant that the health care provider understand the complaint the person is experiencing. That is the first important step thus so as to the appropriate route be able to be taken to find an analysis as well as initiate cure.

Dizziness Reasons

- As organizing faintness into vertigo and lightheadedness types might assist recognize how the stiff workings, occasionally it is meaningful to evaluate usual cause why inhabitants may complain lightheadedness.
- Low blood pressure as a cause of dizziness because of the ions disruptions because of chemicals in food.
- Lightheadedness, dizziness, also the sense of passing out is a general criticism in inhabitants who have low blood pressure. When not sufficient oxygen-rich blood is being transported to the brain, the blood pressure is too low, and its task may be affected. The person may pass out, syncope, if the brain's blood supply is reduced too much. Signs might inferior as a changeable place from lying down otherwise sitting, to standing up.

Symptoms of Dizziness

Following related indications might include, besides feeling faint:

- Shortness of breath.
- Chest pain.
- Vomiting along with nausea.
- Weakness.
- Sweating.

Long-term Effects

A chemical hazard may associate with various long term effects such as liver cancer.

Liver Cancer (Hepatocellular Carcinoma)

The principal liver tumor is a situation so as to take place when typical cells in the liver develop into atypical in manifestation as well as activities. The cancer cells are able to then develop into critical to neighboring normal tissues and may grow both to organs outside of the liver as well as to other areas of the liver.

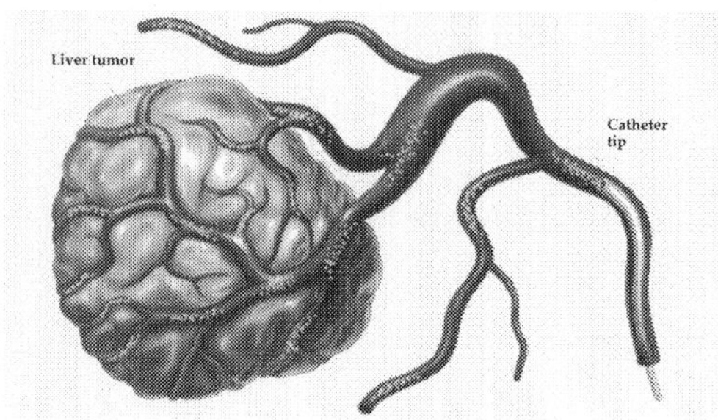

Figure 1.6: Liver Cancer (Hepatocellular Carcinoma)

http://en.isramedic.co.il/images/stories/6c8bbb49faa8f7ba7618b2ca49414f40.jpg

Cancer so as to rise in the channels of the liver is known as cholangiocarcinoma. Cancerous otherwise malignant cells that increase in the normal cells of the liver, hepatocytes are identified as hepatocellular carcinoma.

Symptoms of Liver Cancer

Liver cancer is the cancerous that is the abnormal enlargement otherwise expansion of liver cells frequently hepatocytes in the liver. Supplementary liver cell kinds uncommonly can develop into cancerous excluding the frequently the typical liver cancer is hepatocellular cancer. Other cancer cells can assault the liver for case in point, breast, colon, otherwise, lung cancer cells however these are measured metastatic secondary cancers, not primary liver cancers. Signs of liver cancer comprise upper abdominal pain, weakness, and unplanned weight loss, jaundice, yellowing of the skin otherwise white region of eyes, appetite loss and light otherwise white chalk like stools.

Metastatic Cancer of Liver

In the liver, metastatic malignancy is a situation in which tumor as of other organs has extended all to the liver by the bloodstream. In the body secondary site, metastatic cancer is malignancy that has spread as of the site anywhere it chiefly ongoing the main position to an additional position. At this time the liver cells are not what have become cancerous. The liver has to turn into the place to which cancer so as to start somewhere else has spread. Metastatic cancer has the same type as well as the same name of cancer chamber like an unusual tumor. Cancers so as to extend to the liver are kidney, pancreas, bladder, breast, ovary, colon, lungs, and uterus in addition to stomach is the mainly frequent cancer.

When malignancy starts in the liver, metastatic liver illness is an unusual clause that occurs, primary as well as grows on the way to some more organs, secondary inside body. There are no indications in several people by metastatic malignancy. By means of X-rays, their metastases come in progress otherwise by further testing. Jaundice otherwise inflammation of the liver, skin yellowing might identify malignancy toward liver, has developed.

Liver Cancer Facts

- Liver malignancies by the majority of people who obtain it, hepatic cancer take it in the phase of unremitting, unrelenting ailment of the liver.

- In the United States, incidence rates of hepatocellular cancer are rising because of increasing occurrence of cirrhosis posed by means of nonalcoholic fatty liver illnesses as well as unremitting hepatitis C.
- For liver cancer, there are various cure substitutes. The treatment selected depends upon how much cancer has enlarged as well as the standard physical condition of the liver along with the whole fitness of the patient.

Liver Cancer Causes

- The majority of people who obtain liver malignancy obtain it in the set of unremitting liver ailment long-term liver damage known as cirrhosis, which grows the risk for liver cancer by scaring **the liver. Alcohol use or else abuse,** hepatitis B, in addition to hepatitis C are conditions that are the cause of cirrhosis.
- To environmental, dietary, otherwise lifestyle factors the sources of liver malignancy might be related. As an illustration, researchers at the University of California, in Nov. 2014, School of Medicine San Diego, found so as to long-term contact toward a widespread constituent in soaps as well as detergents that are triclosan, poses cancer in laboratory mice as well as liver fibrosis. Though to cause human liver cancer triclosan has not been proven, it is currently in inspection whether it has harmful healthiness impacts, by means of the FDA to approximate.

Liver Cancer Risk Factors

Is Liver Malignancy Heritable?

In the United States, occurrence rates of the hepatocellular tumor are rising due to rising dominance of cirrhosis posed by means of persistent hepatitis C in addition to nonalcoholic fatty liver illness, steatohepatitis.

By reason of any source, liver cirrhosis is an uncertainty issue for malignancy of liver. In cirrhosis, the risk chances for liver cancer are male, 55 years age or old, Hispanic ethnicity otherwise Asian, in a first-degree effective family record, alcohol use, hepatitis B Moreover C, obesity as well as increased iron content in the blood owing to hemochromatosis.

Persistent hepatitis B disease yet devoid of cirrhosis is an uncertainty aspect intended for liver malignancy.

Liver Cancer Symptoms

- Liver tumor poses no signs of its own. In the right side of the abdomen, it may cause symptoms of pain otherwise a sensation of fullness while eating, as the tumor grows up. Often patients may have to worsen of cirrhosis or chronic liver disease symptoms, which may often precede the growth of liver cancer. For instance, mysterious weight losses, (cachexia) assassination, a decrease in appetite, feet as well as belly increased swelling, swollen legs, as well as eyes and skin yellowing, may be complained by patients.

Diagnoses for Liver Cancer

A physician will organize one the subsequent to verify a diagnosis, once a misgiving of liver cancer rises:

Blood Analysis

Alfa-Fetoprotein (AFP), this may be enhanced in 70% of patients with liver cancer. AFP levels could be normal, in liver cancer. For liver cancer, an arising intensity of AFP is doubtful. In a lot of patients with liver malignancy, des-gamma-carboxy prothrombin surrounded by other labs tests, which may be able to be elevated.

Imaging Learning

MRI and Multiphasic helical CT scan of the liver are the favored imaging for recognition of the location as well as the level of blood delivered to cancer with contrast. If any imaging study is inconclusive, then an alternative imaging study or follow-up imaging study ought to be executed to assist elucidate the treatment. Normally, lesions lesser than 1 cm are hard to distinguish.

- Through surveillance ultrasound of the liver ended each 6 months in a patient by means of an analysis of cirrhosis is the superlative means to identify the liver tumor. It is fine to take care of the liver cancer as rapidly, as it is diagnosed like with primary categories of cancer.
- By a pathologist, liver biopsy is estimated to confirm the imaginary estimation of malignancy of liver executed in the liver to sample tissue by the lesion, in every case which is not necessary, if the lab markers, as well as imaging learning especially, are features meant for liver tumor. Infection, bleeding, otherwise, seeding of the needle track are risks for liver biopsy. Seeding is when cancer cells grow in other areas touched by, for a biopsy, needle cancer cells get on the needle used. The supplementary danger of seeding the liver biopsy pine needle pathway in 1% to 3% of cases carries liver biopsy of theoretical liver malignancy. A reiterate imaging study is suggested at 3- to 6-month time periods if a liver biopsy is unconvincing.

THE SIGNIFICANCE OF CHEMICAL RESIDUALS IN FOOD

There are various advantages and disadvantages of pesticide use by farmers. Pesticides are chemicals that kill pests. To most farmers, pesticide use is an unavoidable side of farming or gardening. This is because pesticides assist in keeping the plants free from pest infestation. Pests are harmful insects whose effects on the crop can be disastrous if they are not controlled. Pesticides assist farmers in preventing crop catastrophe by these harmful insects. Although pesticides are significant to farmers, using them can constitute a risk to the farmer, his family, and animals. However, when used appropriately pesticides may not have a significant side effect on humans and animals. Nevertheless, it is important to suppose the advantages and disadvantages of pesticide use before a farmer starts using them.

One of the major advantages of pesticide use is that they kill pests faster than other pest control techniques. This is because pesticides are certainly formulated chemicals that target specific pests. Once administered in a crop that has been attacked by the pest, pesticides start working suddenly by affecting the normal biological functions of the organs of the insect. Pesticides are also easy to use. With most pesticides, a farmer is just needed to mix the pesticide with a certain amount of water and then spray the crop. There are powder pesticides that a farmer applies to the crop that is invaded by the pests directly. Thus, it takes a few minutes or hours to apply the pesticide and control the pest.

Some of the Advantages and Disadvantages of Pesticides are as follows. So let us check it out some information on advantages and disadvantages to know more about pesticides. Different types of pesticides are available such as herbicides, insecticide, fungicide, nematocide etc., are available in the market.

Advantages of Chemical Residuals in Food

Advantages of Pesticides in Food

- Cost Effectiveness
- Crop Protection
- Control pests
- Greater yields
- Effective and rapid
- Increase food supplies
- Flexibility
- Used to kill weeds
- Prevention of Problems and disease
- Protects Food
- Save human lives

Disadvantages of Chemical Residuals in Food

Disadvantages of Pesticides

- Reduction of beneficial species
- Residues in foods
- Kill natural pest enemies

- Resistance
- Poisoning hazards
- Ground Water Contamination
- Drift of Sprays and Vapors
- Pollutes the environment
- Pesticides can harm wildlife and people.
- Sometimes pesticides may harm various kinds of organisms, which are not pests.
- Some pesticides are persistent.
- Pesticides may accumulate and enter the food chain.
- Health risks to both human as well as animals.

Animal Drug Residues

Animal drug residues may be of concern for foods of animal origin, including muscle meat, eggs, honey, organ meat, fat/skin, and milk. In the United States, animal drugs need approval by FDA before they can be applied to food-producing animals. Depending on the chemical property of the drug, residues of certain drugs can become concentrated during food manufacturing and processing. For example, if a fat-soluble, heat-stable drug residue is found in raw milk, the drug can get concentrated when the milk is converted to full-fat cheese.

Potential effects of drug residues range from short-term effects as a result of critical allergic reactions (e.g., penicillin) to long-term effects from drug resistant bacteria (Dayan, 1993). An example of an unapproved drug residue that has adulterated food is fluoroquinolone, which is an antibiotic that has not been approved for use on honey bees in the United States and has been detected in honey products from certain regions outside the United States (FDA, 2015a).

Drug residues in a food derived from an animal (such as milk) are considered a hazard if a tolerance has not been established for the particular drug-food combination, or if the tolerance level has been extended. Animal drugs used according to labeled directions should not result in residues in meat, poultry, milk, or egg products. When your hazard analysis detects drug residues that require a preventive control, supply-chain controls with a supplier verification program could be an appropriate preventive control to manage the potential risk.

Heavy Metals

Heavy metals, including lead, cadmium, arsenic, and mercury, may be of concern in certain foods as a result of agricultural practices (e.g., use of pesticides containing heavy metals or because crops are grown in soil containing elevated levels of heavy metals due to industrial waste), or the leaching of heavy metals from equipment, containers or utensils that come in contact with foods. Consumption of heavy metals in foods can lead to severe health consequences. For example, lead exposure can impair cognitive development in children (FDA, 2006a).

When your hazard analysis detects a heavy metal that requires a preventive control, the type of control would depend on how the heavy metal could get into your food product. In some cases, high levels of heavy metals may result from the environment (e.g., high lead levels in carrots that were grown in lead-contaminated soil). If your food product contains a food crop that is known to have been contaminated with a heavy metal via contaminated soil, a preventive control such as a supply-chain control with a verification program to ensure that the grower conducts an evaluation of the growing region prior to its use for agriculture may be suitable. In other cases, an unsafe level of a heavy metal such as lead could be introduced into a food product as a result of a food-contact surface constructed with lead solder. CGMP controls, such as the controls on equipment and utensils in 21 CFR 117.40, generally can control chemical hazards such as heavy metals that can leach from food-contact surfaces.

Environmental Contaminants

Environmental contaminants may be of concern in certain foods as a result of their presence in the environment. When your hazard analysis detects an environmental contaminant that requires a preventive control, the type of control would depend on how the environmental contaminant could get into your food product. In some cases, high levels of environmental contaminants (e.g., dioxin) may result from accidental contamination of animal feed (WHO, 2014).

Mycotoxins and Other Natural Toxins

Natural toxins, such as mycotoxins, histamines, and other biogenic amines, and plant-manufactured substances (such as the toxin hypoglycin found in the tropical fruit ackee) are well recognized as hazards in raw or processed agricultural commodities (FDA, 2005a; FDA 2005b; FDA, 2005c; FDA, 2005d).

Mycotoxins are a common group of natural toxins that include aflatoxin, fumonisin, deoxynivalenol (vomitoxin), ochratoxin, and patulin. Mycotoxins are toxic metabolites manufactured by certain molds (i.e., fungi) that may infect and proliferate on agricultural commodities (e.g., grains such as wheat and corn, peanuts, fruits, and tree nuts) in the field and while storage. Mycotoxins may produce different toxicological effects. Some mycotoxins are teratogenic, mutagenic, or carcinogenic in susceptible animal species and are related to various illnesses in domestic animals, livestock, and humans in many places of the world. The occurrence of mycotoxins in animal and human foods is not completely avoidable; minor concentrations of these toxins can be found on agricultural commodities. The occurrence of these toxins on commodities susceptible to mold infestation is affected by environmental factors like humidity, temperature, and the level of rainfall during the pre-harvesting, harvesting, and post-harvesting durations. The molds that produce mycotoxins typically grow and become established in the agricultural commodity during stressful growing and holding situations, such as insect damage to the crop, drought stress, and wet storage (e.g., from condensation). Although mycotoxins are not a hazard requiring a preventive control during times and positions with good growing and harvest situations, a preventive control such as supply-chain controls with a supplier verification program may be suitable if you use agricultural commodities susceptible to mycotoxin formation because growing and harvest conditions alter from year to year.

POTENTIAL RISKS TO FOOD SAFETY

Chemical Carcinogens

More than a century of technologic advances in industrial, agricultural, and commercial products and pharmaceuticals resulted in the vast use and scattering of chemicals. Although chemicals have improved quality of life, few have been adequately tested to identify whether they cause health risks to humans. More than 80,000 industrial chemicals are registered with the USEPA for commercial use in the United States, and there are additional chemicals in cosmetics and personal care products as well as food additives, pharmaceuticals, and pesticides.

Home furnishings used to be made of wood, glass, wool, linen, or cotton, whereas today many are made of plastics, synthetic fabrics, and other products derived from petroleum. Many of these synthetic materials are further treated with other chemicals such as brominated flame retardants, fluorinated water repellants, and agents that modify plastic (such as phthalates and bisphenol A, both of which have estrogenic activity). The USEPA has required extensive testing of fewer than 200 chemicals, leaving producers to determine any possible possibility of the rest.

The IARC lists 107 agents, most of them chemicals, as known human carcinogens (Group 1), 59 agents as probable human carcinogens (Group 2A), and 267 agents as possible human carcinogens (Group 2B). In many cases, Groups 2A and 2B are shown to be carcinogenic in animals, but we have no conclusive evidence as to whether they are also carcinogenic in humans. However, chemicals that pose cancer in animals are likely to also cause cancer in humans. The U.S. Department of Health and Human Services, in the 12th Report on Carcinogens, lists 54 agents as known human carcinogens, and 186 as "reasonably anticipated to be human carcinogens." Although the 186 chemicals noted are not identical to those in the IARC listing, clearly, most are substances that are known or likely to pose human cancer. Many of these chemicals are widely administrated and are usually ingested through water and food and inhaled from the air by large pieces of the population on the regular basis.

The Risk of Exposure to Carcinogens Early in Life

Studies of human populations exposed to ionizing radiation provide the strongest proof that exposure to a carcinogen early in life causes a greater risk than exposure during adulthood. Much of what we know about ionizing radiation and cancer comes from studies of atomic bomb survivors. Those children who were younger than age 5 at the time of the blasts had more than twice the relative risk of developing cancer later in life compared with older children who were between the ages of 5 and 15 at the time of the outburst. The latency period for cancer development was short in some cases, but the people who outlive the nuclear explosions in Hiroshima and Nagasaki in 1945 are still at elevated insecurity today. Exposure to the radiation enhanced risk for all kinds of cancers. The relative risk for development of breast cancer after the atomic bombs was much higher for girls who were prepubertal at the time of the blasts than for women who were older than 40 and the degree of probability for other solid tumors also change with age, being highest early in life.

There is other evidence that early life exposure to radiation extends risk for cancer later in life. Prepubertal children exposed to ionizing radiation from atomic bomb tests in the Marshall Islands from 1946 to 1958 developed thyroid cancer 15 to 29 years after exposure, but exposed adults had little or no extended risk. Similarly, after the accident at the Chernobyl nuclear reactor in 1986, an increased risk for thyroid cancer was found among individuals who were younger than 18 at the time of outburst, and their risk lasted elevated for more than 2 decades. The risk for thyroid cancer did not increase among adults. Sadetzki et al found that thyroid cancer after childhood exposure to ionizing radiation for tinea capitis was positively related with dose and negatively associated with age at exposure. The risk was enhanced 10 to 19 years after exposure, peaking at 20 to 30 years afterward, and decreasing 40 years afterward. Somewhat same results were found for risk for skin cancer after sunburn: exposure when 8 to 12 years of age produced the strongest increase in risk. In today's world, children are often exposed to other sources of radiation, including background radiation, radiation from medical treatments, radon, and ultraviolet radiation from the sun. Of specifically concern is children living in areas with high concentrations of radon and exposed to frequent computed tomography scans.

Diethylstilbestrol (DES) is another example of a chemical that places those exposed during gestation at higher risk than the risk for those exposed upcoming in life. DES is a synthetic and potent estrogen that was widely given to pregnant women from the 1940s to 1971, with the false assumption that it would decrease the risk of spontaneous abortion. Both daughters and sons of women who took DES show a variety of severe developmental and health effects. Women exposed in utero to DES were at enhanced risk for clear cell adenocarcinoma of the vagina and uterus when they attained late adolescence and suffered another peak in risk for cancer (this time, breast cancer) after age 40. There are most of other chemicals that have estrogenic or other hormonal disruptive activity. Endocrine-disrupting chemicals have the potential to change many hormonal systems, and new results show that many effects occur at low doses and show nonmonotonic dose-response curves, which complicates risk appraisal. Many carcinogenic exposures are associated with epigenetic variations as well, which may contribute to cancer in future generations.

Studies of human exposure to chemical carcinogens during early life are few. However, animal studies, reviewed by Ginsberg, show that critical exposure of juvenile animals to eight various carcinogens resulted in at least a twofold greater sensitivity than a similar risky exposure to adult animals. This finding applied to cancers of the liver, breast, blood, lung, kidney, and nervous system. Some chemicals, notably benzopyrene, showed a ninefold greater chance for liver cancer when administered to neonatal animals than when administered to adult animals. Barton et al reviewed animal studies on mutagenic carcinogens and found a 10-fold greater susceptibility resulting from perinatal exposure than from adult exposure.

The U.S. Environmental Protection Agency (USEPA) identifies the unique vulnerability of children to carcinogens and has incorporated a 10-fold risk chance factor for exposure to known carcinogens of children younger than 2 years, a threefold adjustment factor for children and adolescents ages 2–15, and no adjustment for anyone older than 15 years. These adjustment factors are to acknowledge the greater susceptibility of younger children. Another significant variable is the mechanism of action of different carcinogens. USEPA supposes that carcinogenic chemicals have a linear dose-response relationship, which is to say that for a mutagenic carcinogen there is an enhanced risk at any level. Thus for mutagenic chemicals and ionizing radiation, there is no level of exposure that is "safe". There is debate as to whether the similar principle applies to nonmutagenic carcinogens, and efforts to develop a framework for understanding and awareness about cancer by mode of action are under way.

Human Carcinogens to Which Young People Are Exposed

10 common carcinogens as examples of chemicals to which children and adolescents are exposed daily are explained below. These are by no means the only carcinogens of notice, but they illustrate the ubiquity of exposure. Exposure through food is the specifically unique route, especially for lipophilic chemicals present in animal fats. The most catastrophic of these are bioaccumulative and persistent (e.g., dioxins, furans, PCBs, chlorinated pesticides). These are known or probable human carcinogens and are present essentially in all animal fats. PAHs are also ubiquitous in food, air, and water resulting in exposures via inhalation and ingestion. Many carcinogens are ingested through drinking water and others are inhaled. Indoor air is a major source of carcinogenic volatile organics.

Dioxins and furans are products of combustion and backyard burning of the garden and domestic wastes and were also inadvertent byproducts of some industrial processes, such as the manufacture of Agent Orange, used as an herbicide in Vietnam. The major source in the United States used to be municipal and medical incinerators because dioxins are produced when plastics or other chlorine-containing substances are burned.

Thanks to the Clean Air Act of 1970, and its amended forms in 1977 and 1990, continued dioxin emissions into the air have been greatly minimized, but dioxins formed earlier are still in the environment and in our food. Recently in the United States, the greatest source of dioxin production is backyard barrel burning, followed by medical waste incineration. PCBs are industrial chemicals whose manufacture was discontinued in 1977 because they are persistent both in the environment and the human body because the chlorines make the molecule very stable. Both dioxins and furans (with 75 congeners each) and PCBs (with 209 congeners) are chemicals with different numbers of chlorine atoms around benzene carbon rings. One dioxin, one furan, and one PCB congener are rated by IARC were known as human carcinogens, and the remaining dioxins, furans, and PCBs are rated as probable human carcinogens. All are lipophilic, stored in human body fat, and vastly distributed in nearly all animal fats.

In 2003, the Institute of Medicine released a report entitled Dioxin and Dioxin-Like Compounds in the Food Supply: Strategies to Decrease

Exposure. This report documented the vast distribution of these chemicals, even though dioxin release and PCB manufacture had been prohibited for 30 years. This document reported results as of 2001 from an ongoing study called "The Total Diet Study: Dioxin/Furan Survey" conducted by the Food and Drug Administration (FDA). The survey was a random sampling of supermarket foods that were measured for dioxin-like activity. All dioxins, furans, and a subset of PCBs act through a usual cytoplasmic receptor, called the aryl hydrocarbon receptor. Their common mode of action led the World Health Organization to develop the idea of toxic equivalents (TEQ). This measures overall dioxin-like activity for all of these substances, comparing them against the most potent one, 2,3,7,8-tetrachloro di-benzo-p-dioxin.

Table 1.1: Dioxin equivalent amounts of common foods consumed by children and adolescents

Food category	Food item description	Mean TEQ (parts per trillion) nondetects = .05 LOD
Dairy	Milk, whole	.0110
	Cheese, cheddar	.02534
	Cheese, swiss	.0833
	Ice cream, vanilla	.0401
Fats and oils	Butter	.02202
	Olive oil	.0641
Meats	Beef, chuck roast	.04595
	Frankfurter, beef	.01983
	Hamburger, fast food	.0678
	Cheeseburger, fast food	.0739
	Pork bacon	.02116
	Pork sausage	.01962
	Bologna, luncheon meat	.01558
Poultry	Fried chicken, fast food	.0550
	Turkey breast, roasted	.0538
Fish	Fish sticks	.0335
	Tuna, canned in oil	.0838

Cereals and bread	Oat rings	.0224
	Corn flakes	.0106
	Crisped Rice	.0157
	Oatmeal, quick cooked	.0063
	Bread, white	.0071
Prepared foods	Lasagna with meat	.0408
	Macaroni and cheese, box	.0242
	Pizza, pepperoni, carry out	.0205
	Taco or Tostada, carry out	.0619
	Peanut butter	.0068
Snacks	Brownies	.0151
	Cake, yellow, white icing	.0379
	Chocolate chip cookies	.0206
	Popcorn, popped in oil	.0411
	Candy bar, milk chocolate	.0222
	Potato chips	.0263
Vegetables and fruit	Bell pepper, green, raw	.0039
	Broccoli, boiled	.0035
	Carrots, fresh, boiled	.0035
	Lettuce, iceberg, raw	.0043
	Peas, green, boiled	.0020
	Potatoes, French fries, frozen, heated	.0614
	Apple, red, raw	.0065
	Applesauce, bottled	.0055
	Orange, raw	.0073
	Watermelon, raw	.0046
Condiments	Mayonnaise	.0136
	Mustard, yellow	.0079
	Tomato catsup	.0035

LOD = limit of detection; TEQ = toxic equivalents.

Table 1.2: Estimated daily intake of dioxin-like compounds from food by consumers of high versus low combined amounts of meat, poultry, and fish

Population	Total food intake/ day gm/ kg (body weight) per day	Consumption of meat, poultry, fish	n	Dioxin toxic equivalent intake pg/kg body weight detects = .5 (LOD)
Females and Males ages 1–5 years (not breastfeeding)	90.5	High	5,306	1.76
		Low	1,035	1.26
Males and females ages 6–11 years	53.4	High	1,735	1.14
		Low	166	.077
Males only ages 12–19 years	38.6	High	675	.089
		Low	18	.047
Females only ages 12–19 years	30.6	High	615	.069
		Low	71	.041
Males only ages 20 +	32.3	High	4,608	.069
		Low	110	.040
Females only ages 20 + (not pregnant or lactating)	29.3	High	4,128	.059
		Low	281	.038

LOD = limit of detection.

A .05 LOD means half of the lowest concentration at which a chemical can be detected above "background noise."

Low combined meat, poultry, and fish ≤ 1 oz (29 g)/2 days.

Dioxins are found in dairy products, eggs, meat, and fish, and dietary intake is believed to contribute up to 90% of exposure, with intake in children higher than in adults on a body weight basis.

Although it's impossible, to totally avoid exposure to carcinogens, there are some notable steps that can be taken. This has significant importance during pregnancy. Selecting foods that have lower concentrations of carcinogens is one obvious step. Choices in use of personal care products, avoiding extensive exposure to radiation and tobacco products are others.

Most cancers are a consequence of exposure to environmental carcinogens, many of which are chemicals that get into the human body by food, water, air, or absorption through the skin. Genetics is important in determining individual susceptibility to cancer, but few cancers are due solely to heritable genetic factors. We must find methods to minimize human exposure to carcinogenic chemicals and ionizing radiation. Specifically important is finding ways to reduce exposure during gestation, childhood, and adolescence because exposure during these critical durations of development leads to cancers later in life.

SEVERAL FOOD SCANDALS

Since the 1980s: Mad cow disease.

1981: "Toxic oil syndrome": Hundreds of people passed away by the consumption of rapeseed oil denatured with aniline.

1985: Up gradation of wine to table-wine quality by the addition of ethylene glycol which is a frost protection agent.

1985: Pasta products made by usage of spoil eggs.

1994: Chilli powder contamination with lead tetroxide.

1996: Synthetic "apple juice" concentration.

1999: Dioxins in the feed.

2001: Pork contamination by hormones, vaccines, and antibiotics.

2002: From China, antibiotics in honey.

2003: From Eastern Europe, adulterated wine prepared (with water, sugar; added alcohol, and coloring).

2004: Banned dyes usage in spices.

2013: In Europe, meat adulteration scandal; undeclared horse meat were found in foods advertised as containing beef, in some cases, as much as 100% of the meat content.

Food Scandals in V 4 Countries

Poland: Rodent poison cause contamination in milk powder, Salmonella, infected chicken meat, road salt in food, antibiotics in honey, methanol in alcoholic beverages (more than 30 deaths).

Hungary: colored pork, and sold as beef.

Slovakia: yogurt with glass.

CHAPTER

2

SURFACE TO FOOD: EXPOSURE WITH CHEMICALS

Food is able to eat after the production, processing and sometimes after cooking properly and by all these methods food comes in contact with different chemicals.

CHEMICALS MIGRATION FROM PACKAGING TO FOOD

Normally, primary packaging incident where food been imputed by the relocation of chemicals have concerned packaging in direct contact with foodstuff. Transfer can also occur via evaporation and then leach into food via the gaseous phase.

SELECTION OF FOOD PACKAGING MATERIAL, MIGRATION AND ITS EFFECTS ON FOOD QUALITY

ABSTRACT

Technologies for different food products, the shelf life of food is increased, food quality and freshness can be preserved by selecting the suitable packaging material. Food packaging is one of the stages of food production that enables foods to reach consumers carefully. If food packaging material is not selected properly food may be contaminated by chemicals, while basic function of food packaging is to protect food from contaminants.

Migration, under specific conditions, can be explained like mass transfer of chemicals from food packaging to food while storage as well as usage. The packages are evaluated as main contamination source of chemicals because food may be kept in packaging for a long time. Furthermore, because migrating substances generate remains on the food they unfavorably affect food safety and quality. These materials passing from packaging to food can cause flavor loss and significant health problems and taste. It is an obligatory requirement for consumers to decide the impact of packaging in contact with food on food safety as the last chain in food safety food. Substances migrating to foodstuff give an awful odor to provisions and as a result, reduce customer option.

Today, with different packaging technologies, new packaging materials are produced. Customers may not use the packing material as projected constantly this condition may lead to different exposure. In terms of food safety, each newly produced packaging material must be evaluated. It is necessary to take into account that packaging materials in contact with food should not give its ingredients to food while querying the packaging of foodstuffs we buy as consumers. Food quality affects human health as well as human health alters the quality of existence. It is essential to be alert of the likelihood of chemical immigration in food production, distribution, sales, as well as consumption stages and it must be diminished.

GENERAL DESCRIPTION

After packaging in one way or another all industrial products produced by industry are forwarded to consumers. Therefore to preserve as well as protect the product contained within packaging is the basic function of packaging.

Food packaging is in contact with the food depending on conditions like packaging material as well as foodstuff properties, packaging as well as storage temperatures, ultraviolet light exposure as well as storage duration of the product. Packaging protects, maintains, presents, promotes as well as displays product all through transport, storage and use economically and susceptible to surroundings throughout the whole life period of products.

Migration to food from packaging material may also take place during migration from the packaging material to food. This contact is one of the significant causes so as to affect quality, appearance in addition to the shelf life of food. This contact affects the permeability of the packaging, as well as guides, increase the rate of reactions like oxidation as well as flavor transition. Such interactions might enhance the danger of exodus.

Mass transfer of parts from amounts of high levels to areas of the low amount will take place inside the food as well as within the packaging material diffusion is caused by concentration gradients. Moreover keeping foods harmless from contamination as well as keeping hold of the nutritional characteristics Furthermore sensory properties of foodstuff, wrapping makes available extra characteristics so as to be significant to customers.

FOOD PACKAGING AND SAFETY

Food safety is described as conformity with necessary needs while processing, production, storage, transport as well as circulation of food as well as taking vital capacity to ensure healthy food formation. In comparable to developing economy fast food consumption, for that reason, fast food production is increasing. Ensuring the safety of materials in contact with food from field to table as the final relationship of food safety chain, evaluation of its result on food quality as well as safety plus recognition of health risks resulting from these materials are obligatory for the defense of constitutional rights. Recently, food packaging has gained a wider significance in food safety owing to the chance of migration of chemicals from victuals exposed materials. The term 'migration' usually generalize a distribution process, which might be strongly influenced by a contact of the wrapping material with the foodstuff.

MIGRATION

As a result of migration occurring between food and food packages; required characteristics of packaging materials may change, loss of quality can occur in food ingredients or food material can be degraded. Therefore, migration is very important in food packaging as well as a range of suitable packaging materials for foods as well as storage circumstances must be measured.

Factors Affecting the Migration

There are many values that affect the rate as well as the speed of migration from materials exposure by food to food. These limitations are able to be summed up as follows direct otherwise indirect exposure of packaging substance by food.

For plastics chemical properties vapor pressure, polarity, molecular size as well as structure, etc. are features of the materials in contact with food thickness, permeability, etc. of the migrant. Contact time, as well as temperature, is initial concentration of migrant in the packaging material.

Food or Stimulants Components in Contact with the Packaging Material

Commonly minute molecules roam faster than larger ones. As migration increases significantly with increasing temperature it takes place through physicochemical methods. Migration depends on scheduled time. At the beginning of shelf- life a food packaging companionable might be incompatible at the closing stages of shelf-life (e.g., long-lasting products). Fatty foods will tend to attract nonpolar immigrants although water-based foods are likely to magnetize pole immigrants. Food with a high surface area based on its volume for example high surface area - pasta is usually more susceptible to transportation. Different permeability potential is against migrants' different surface.

It is supposed that glass and metal completely forbid migration. Though, migration into the metal containers into food should be considered in addition to minimizing covering. Another factor to be considered in concerning covering practice, food consumption, as well as migration data, is the ratio of surface area of the packaging to the volume of foodstuff creation. Consequently, the sizes, as well as the set-up of the packages are too very significant.

Migration Types

In a number of different ways migration may take place:

Contact Migration

It entails the food contact faces of the packaging into the food from the direct transfer of substances. Instances may include the transport of substances from a cardboard pizza box to the underneath of a pizza, or transport of substances from a plastic tub, tray, pouch otherwise wrapping into food.

Gas Phase Migration

It contains the shifting of volatile substances throughout the airspace among food as well as packaging and into the foodstuff by the procedure of dispersal. Mineral oil from recycled paperboard, into solid dry foodstuffs, might be a good example of the diffusion.

From the cartons, mineral oil may potentially migrate, through the airspace, through a plastic inner pouch subject to its barrier characteristics as well as throughout a second airspace into the foodstuff.

Set-Off Migration

It involves the transfer of substances from printing inks, coatings or varnishes from the printed, not food contact side of the packaging to the food contact side, as a result of the stacking of printed items e.g. carton flats, or when winding a printed film into a reel.

Condensation or Distillation Migration

It involves the transfer of substances into food during heating processes for example sterilization or boiling of pouched food etc., or the cooking of food in an oven/microwave in cartons or trays.

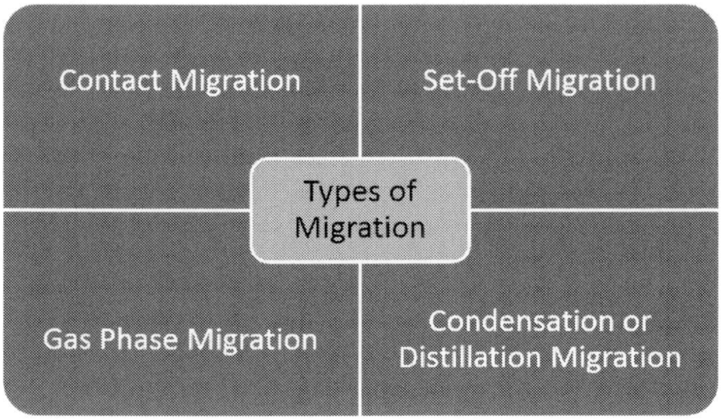

Figure 2.1: Types of Migration.

Migration in Plastic Based Packaging Materials

There are more than 30 types of plastic materials including widely used polyethylene, polypropylene, polycarbonate and polyvinyl chloride and chemicals are known to migrate to food among these materials raise concern about health.

Chemical migration is more likely to occur for materials such as plastics, elastomers, paper, and board. Decomposition products from additives or monomers will also migrate into the food under proper conditions.

Migration in Paper Based Packaging Materials

Paper and cardboard materials are used for many different purposes especially like the packaging of hot drinks and food in restaurants described as fast food, in tea bags, filter papers, oven/bakery products and dry foods. In addition, they are used for packaging of foods such as flour, sugar, and pasta and in the transport of vegetables, fruit, and eggs frequently. Due to chemical substances arising from their natural structures and compounds added during production and in case they are used again paper and cardboard materials contain a large number of toxic organic molecules. Compared to other materials such as glass, metals, and plastics in contact with food, paper/cardboard materials may keep especially volatile compounds that cause bad smells due to their porous structure.

To slow the commencement of oxidative deprivation of synthetics by an introduction to UV radiance, antioxidants are added up in the direction of polymer resins diversity. Polymers degrade by reason of the exploit of extremely imprudent free radicals produced by means of radiation, heat as well as perfunctory cut off so as to be frequently increased by means of the metallic contamination occurrence. In foodstuff covering, at the elevated temperatures oxidation level rise, counting exposure to the hot foodstuff otherwise introduction to infrared heating, rejoinder dispensation as well as potentially, MW heating.

Metal Based Packaging and Migration

Tin-coated metal cans act as a barrier against gas, water vapor and flavor like glass. It is opaque and light proof and recycling are possible. When steel and aluminum are used, it is required to make lacquering to prevent food interaction because these materials are not inert as glass. The migration from metal boxes arises from metals they contain and initial components or derivatives of polymers used in lacquering.

Chemicals in the structure aluminum containers and/ or added during their production can migrate to foods and can reach the size to be harmful to human health. Sources of Al transition are contacted environment during production and before submitted to consumption, packaging, kitchen materials, cutting and cooking tools and many materials. Although contact time with the manufacturing and packaging apparatus is less; foods can stay for long periods in their packaging during shelf life. There are studies are carried out to limit overall and specific migration values in some countries for packaging group. However, there is no common legislation in EU and in our country with regard to materials like metal, silicon, paper and so.

This poses a risk to human health. For the case of the beverage cans made from tinplate, the control of the migration of iron acquires an even greater importance; because extremely small levels of iron migrated to the drink (0.5 ppm) can already compromise the flavor of the drink.

Migration and Health Effects

Protection of food packaging /materials in contact food during storing or handling food and prevention of migration of any unwanted nonfood components to food are main tasks of food 1st International Conference on Quality of Life June 2016 205 packaging. In the compositions of materials in contact with food according to feature given to food: they can be monomer, starting materials, catalyst, solvent and additives (antioxidants, antistatics, plasticizers, heat stabilizers, dyes, and pigments) and they may be also known and unknown mixtures of these substances, impurities, reactions and decomposition products. Every day, toxic effects of chemicals used in non-food materials in contact with food emerge and mostly affected groups are especially babies and children consumer groups.

In addition, specific migrant can sometimes transform into toxic metabolites migrant shape and can also be difficult to identify. Consumers don't use packaging materials always as offered and this causes different exposures.

Diffusing agents can affect the organoleptic properties of the food and health of humans negatively. Many plasticizers and additives are described as Endocrine Disrupting. They adversely affect the human reproductive system and are carcinogenic. Some of these chemicals are phthalate esters, alkyl phenols (APs) and 2,2- bis (4- hydroxyphenyl) propane known as bisphenol A, its derivatives and di (2-Ethylhexyl) adipate (DEHA). Repeated exposures to even very low concentrations of these components create serious toxic effects. Therefore, as much as intake, food packaging material should also be safe.

Furthermore, the list of hazardous monomers, oligomers, and additives continues to augment in order to ensure that the consumer safety is in current agreement with the HACCP, which is continuously gaining ground. Structure instruction 89/109/EEC (CEC 1989) supplies 2 fundamental main beliefs to which the law is founded. These comprise "inertness" of the food exposure resources as well as editorial, as well as "safety." The code of motionlessness is so as to any article, material, otherwise it works ought to be still sufficient not to cause several fitness risks, undesirable variations in provisions deterioration otherwise composition of foodstuff character. The utilization of elevated amount of PS foodstuff covering might cause fitness unease as remaining styrene be able to transfer as of PS. BPA, Epoxy resins, as well called 'bisphenol A diglyceride ether (BADGE)', encompass cytotoxic sound effects in existing tissues, as well as have been exposed to enhance the speed of separation of the unit cell. Consistent with Commission directive (EU) No 10/2011, in PVC plastic, the vinyl chloride monomers are able to cause severe toxicity within the human being. In polyurethane polymers, the isocyanates consumed as well as bonding agents take a little menace of oral toxicity, however an elevated danger of toxicity as of inhalation otherwise dermal introduction. BADGE is a monomer as well as the chief constituent of epoxy resins in favor of inside container coating. Unreacted BPA within the synthetic inside layer of the containers otherwise cans be able to transfer into foodstuff while storage as well as heating.

BADGE is added to the polymers to serve up like an antioxidant, however, might have unprocessed BPA. It may not be damaging to organisms if the concentrations of materials are less than the particular maximum value. Both Monomers, as well as oligomers, are liable to transfer as of wrapping materials into the foodstuff. Severe fitness dangers might happen as the quantity unprocessed monomers otherwise low-molecular-weight materials in foodstuff arrives at a particular maximum value as well as so immersed to the human being (EU 10/2011). The restraints have moreover been functional to the SML for several meals as well as primary aromatic amines transfers as of synthetic parcels into foodstuff (EU 10/2011). A plasticizer in PVC, 'DEHA (di-(Ethylhexyl) adipate)', is able to transfer as of wrapping to fatty foodstuff, as well as the introduction is controlled to maintain the total daily intake (TDI) under 0.3 mg/kg body weight. Gao as well as others considered the exodus of 8 antioxidants: BHT, BHA, Irganox 1035, Cyanox 2246, Irganox 1330, Irganox 1076, Irgafos 168, Irganox 1010, as well as its dreadful conditions item for consumption DBP at 40 °C in a 10-d storage learning DBP, BHA, Cyanox 2246, Irganox 1035, BHT moved into aqueous stimulants: Irganox 1330, BHT as well as Irganox 1010 were noticed inside oil.

Beldì as well as others considered the impact of fat concentration and storage 206 A., İçöz, B., Eker warmth on the exodus of Irganox 1076 as of LDPE toward a number of foodstuff, chicken, cheese sauce, margarine, chocolate, milk, mayonnaise, soft cheese, orange juice, salmon, pork as well as wheat flour in addition to provisions stimulants; 3% acetic acid, distilled water, rectified olive oil, ethanol 10%, 95% ethanol, as well as isooctane. They instituted so as to immigration enlarged when the chubby substance of the storage as well as food, temperature improved, with the chief intensity of exodus '1413 μg/dm2' meant for chocolate '32.1% fat' by '40 °C' later than 30 d of storage.

Linssen, as well as others, establish so as to exodus of antioxidants 'Irgafos168 as well as Irganox 1076' enhanced with rising content '40% to 100%' of ethanol within foodstuff stimulants, with the uppermost percentage of immigration in '100% ethanolic' stimulants.

Significant appliances of PS comprise packaging trays for cheese along with yogurt, cups, and so on. Quite a lot of learnings account the exodus of styrene into foodstuff as well as approximate the daily styrene introduction of 18.2 to 55.2 μg meant for organisms, with an annual introduction of 6.7 to 20.2 mg. This intensity of introduction poses frustration of the human skin and organs, as well as neurological disruptions.

Likely as well as others considered the relocation of styrene as of diverse food-contact PS foam supplies, egg cartons, meat trays, plates, hinged in addition to cups accomplish container, to the oil, combination of sunflower, vegetable, as well as canola oil, in addition to '8% ethanol by 70 °F, 21 °C meant for 10 d, 120 '49 °C', as well as 150 °F '65.5 °C' in favor of 1, 4, as well as 10 d.

Exodus enhanced as of 1 to 10 d as well as originated to be comparative to the square root of the enlarge in time on a definite warmth in favor of the entire commentaries apart from used for drink cups. Movement of chemical complexes as of wrapping polymers into foodstuff must be estimated to make sure so as to the content of migrating apparatus assemble conformity principles put through regulatory agencies. As temperature enhances, the distributions of monomer, oligomer, as well as additional composite boosts, as well as it may consequence in elevated distribution otherwise migration rates as of wrapping substances. The dissemination coefficient of constituents of covering substance enhances 6- to fold up, as wrappings may be uncovered to severe temperature changes, for illustration, as of freezer temperature to food preparation warmth.

Microwave food packaging has become a tremendous element in the food manufacturing process. The purpose of any food packaging regulations concerned with the safety of the food is to control and limit the migration of substances from the packaging into the food.

Monomers from plastic microwave packaging, chemical compounds like additives or catalysts may filter (leach) from food containers and can cause bad smell, taste and even cause health risks and harmful effects. Generally, when food is heated in a microwave oven with plastic packing it has been shown that there may be massive migration. And also food with high water content may migrate to oil and liquid foods.

Chemical substances so as to be included within the packaging of polymeric substances to develop functionality might interact by means of foodstuff constituents while handing out or else storage as well as transfer in the foodstuff. One time these chemical substances achieve a particular frontier, foodstuff safety, as well as quality, might be endangered.

Potential chemical immigrants comprise antioxidants, monomers, slip compounds, thermal stabilizers, as well as plasticizers.

Migration is an important factor affecting the quality and shelf life of food and it also affects the human health. In order to protect the quality and safety of food, it is necessary to be careful about properties of selected packaging materials. Conditions that could lead to migration must be known and migration capabilities of the new technology packaging material must be determined.

More research should be conducted about chemicals that cause migration and effects on human health.

CONCLUSIONS

Migrants can arise from classic migration or from degradation of the packaging material. Currently, only plastics have EU-wide regulation. Specific chemical-by-chemical regulation of all migrants in food not effective. Potentially 10,000s of migrant chemicals may be found in food. Detailed scientific assessment of all would be an enormous task.

It may involve danger to human health, the components of packaging materials must not exceed the food, the food should not result in changes in the composition and the organoleptic properties of food should not change.

CHAPTER

3

EXPOSURE OF FOOD WITH ARTIFICIAL CHEMICAL RESIDUALS

Food can come in contact with artificial chemical residuals by many different ways. Following are some chemicals explained below:

FERTILIZERS

Mechanical handing out of earth so as to it is within the appropriate substantial situation for planting is typically recommended to as cultivating; adding trace elements as well as nutrients is known as fertilizing. Equally both processes are important in farming operations.

Related Explanation about Fertilizers

Harvesting

Harvesting Machinery

For cutting cereal grains and threshers for separating the seed from the plant, harvesting machinery is normally classified by crop reapers. The supplementary modern combine cuts, threshes, as well as cleans the grain in one action. By mechanical corn pickers, corn maize harvesting is performed that break the ears from the stalk so that only the grains, as well as cobs, are harvested. After or with picking, corn shelling may be done mechanically in the field.

Effects of Fertilizers

Fertilizers leave several effects on the ecosystem including humans, animals as well as non-livings.

Rainmaking

Attempts to raise the amount of precipitation from clouds by seeding them with salt otherwise silver iodide have been made for almost three decades. Beyond the means of an individual farmer, both aircraft, as well as ground generators have been engaged, but the techniques are typical. Consequences recommend so as to cloud alteration is completely feasible, but the proof of enhanced rainfall at a concentration of statistical importance is a difficult disturbance. Under atmospheric conditions, success has been greatest where natural rainfall is mainly possible. However, the view of modifying winter clouds to enhance snowfall in mountain area appears to be somewhat more promising.

Pollution Effects

Air has physical characteristics as well as a chemical composition so as to be vital parameters of life for both plants as well as animals. Temperature, water vapor, movement, oxygen, as well as carbon dioxide in the air have a direct impact on food in addition to fiber fabrication. By introduction of contaminants into it air quality is altered, as well as agricultural activities consuming such air might be affected harmfully. By air pollutants, damage to plants is related to meteorological circumstances, chiefly temperature inversions in the air.

Agricultural Damage by Air Pollution

Burning coal, as well as petroleum, fabricates sulfur oxides. For over a century air pollution has exaggerated agriculture. Fluoride effects as of smelting, glass as well as ceramic forms. Rising amounts of ammonia, chlorine, ethylene, mercaptans, carbon monoxide and nitrogen oxides are present in the air. Motor vehicles, as well as growing population, make photochemical air pollution affecting not just the urban levels but as well the contiguous rural regions.

By the subsequent factors, the effects of air pollution on plants as well as animals may be measured:

(1) Enzyme systems interference
(2) Modification in cellular chemical constituents as well as physical structure.
(3) Retardation of expansion as well as minimized production due to metabolic changes.
(4) Chronic, instant tissue deterioration.

Pollutants that enter the air from sources other than agriculture, as well as that manufacture plant response, are classified as:

(1) Acid gases.
(2) Products of reaction results in the air.
(3) Various effluents.

Acid Gases

Acid gases contain fluorides, sulfur dioxide, as well as chlorine. Some plants are injured by contact with less than one part per billion, hydrogen fluoride as it is extremely toxic to plants. The damage apparently occurs initially to the chlorophyll, producing a mottled chlorosis as well as later on the assassination of the cells. Usually, the plants so as to accrue fluoride eagerly are the most tolerant plants vary in degree of forbearance to hydrogen fluoride. Corn is complementarily vulnerable than tomato. During periods of rapid growth, all plants are most susceptible to fluoride injury.

PRODUCTS OF COMBUSTION

The primary products of combustion include ethylene, acetylene, propylene, and carbon monoxide. Ethylene is known to adversely affect the plants. The presence of the other combustion products may also do the same; however, the higher concentrations (more than the typical amount present in the polluted air) of them would be required to adversely affect the plants.

Air Pollution by Agriculture

The introduction of agricultural technology also leads to contribute air pollution because of the use of pesticides, odors, smoke, dust, allergenic pollens, and trash etc. The main concern of the public about pesticides makes it imperative for the agriculture industry to carefully control the pesticide technology and to look for and employ the alternative eco-friendly methods.

Soil and Water Pollution

Pollutants from Agriculture

Some pollutants from agriculture have adverse influences on agriculture itself, as excesses of plant nutrients and salts from irrigation. These pollutants and others also affect the environment at large.

Salinity

Salinity is a major problem in irrigation agriculture. Through evapotranspiration, salts in the irrigation water become more concentrated in the drainage effluent. It is therefore claimed that water quality is seriously impaired by irrigation agriculture. Irrigation water always contains some salt, most of which is excluded by the plant roots; since the evaporated water is pure, the soil accumulates the residual salt, which is added to what the arid soils already have in abundance. This accumulation of salt must be eradicated if plants are to be grown at all, and it is removed by leaching with excess water.

FOOD AND COLOR ADDITIVES

There are thousands of ingredients used to make foods. The Food and Drug Administration (FDA) maintains a list of over 3000 ingredients in its data base "Everything Added to Food in the United States", many of which we use at home every day (e.g., sugar, baking soda, salt, vanilla, yeast, spices, and colors).

Figure 3.1: Bread having food and color additives.

http://img.aws.livestrongcdn.com/ls-article-image-400/ds-photo/getty/article/251/55/87971550_XS.jpg

Food and Color Ingredients Added to Food

Consumers often take for granted many useful functions that additive may perform while added in food. Some additives could be eradicated if we were willing to grow our own food, harvest and grind it, spend many hours cooking and canning or accept increased risks of food spoilage.

Following are certain reasons why ingredients are added to foods:

- To Maintain or Improve Safety and Freshness
- To Improve or Maintain Nutritional Value
- Improve Taste, Texture, and Appearance

Food Additive

In its broadest sense, a food additive is any substance added to food. Legally, the term refers to "any substance the intended use of which results or may reasonably be expected to result -- directly or indirectly -- in its becoming a component or otherwise affecting the characteristics of any food." This definition includes any substance used in the production, processing, treatment, packaging, transportation or storage of food.

Figure 3.2: Food Additives.

http://www.bdwelfaresociety.org/wp-content/uploads/2015/11/Food-additives-edited-unique-pic-02..jpg

Direct food additives are those that are added to a food for a specific purpose in that food. For example, xanthan gum -- used in salad dressings, chocolate milk, bakery fillings and other foods to add texture, are the direct additive.

Indirect food additives are those that become part of the food in trace amounts due to its packaging, storage or another handling. For instance, minute levels of packaging substances may find their way into foods during storage.

Color Additive

A color additive is any dye, pigment or substance which when added or applied to a food, drug or cosmetic, or to the human body, is able (alone or through reactions with other substances) of imparting color.

Color additives are used in foods for many reasons:

1) To offset color loss due to exposure to light, air, temperature extremes, moisture and storage conditions.
2) To correct natural variations in color.
3) To enhance colors that occur naturally.

Figure 3.3: Color additives in ice cream.

https://upload.wikimedia.org/wikipedia/commons/3/31/Ice_Cream_dessert_02.jpg

4) To provide color to colorless and "fun" foods. Without color additives, colas wouldn't be brown, margarine wouldn't be yellow and mint ice cream wouldn't be green. Color additives are now recognized as an important part of practically all processed foods we eat.

Figure 3.4: Food and color additives in food.

Additives Approval for Use in Foods

To market, a new food or a color additive (or before using an additive already approved for one use in another manner not yet approved), a manufacturer or other sponsor must first petition FDA for its approval. These petitions must provide evidence that the substance is safe for the ways in which it will be used.

GROUP I - Prior-sanctioned substances - are substances that FDA or USDA had determined safe for use in food prior to the 1958 amendment. Examples are sodium nitrite and potassium nitrite used to preserve luncheon meats.

GROUP II - GRAS (generally recognized as safe) ingredients - are those that are generally recognized by experts as safe, based on their extensive history of use in food before 1958 or based on published scientific evidence. Among the several hundred, GRAS substances are salt, sugar, spices, vitamins and monosodium glutamate (MSG). Manufacturers may also request that FDA review the industry's determination of GRAS Status.

When evaluating the safety of a substance and whether it should be approved, FDA considers:

1) The composition and properties of the substance.
2) The amount that would typically be consumed.
3) Immediate and long-term health effects.
4) Various safety factors. The evaluation estimates an appropriate level of use that includes a built-in safety margin - a factor that allows for uncertainty about the levels of consumption that are expected to be harmless.

Review

Food and color additives are strictly studied, regulated and monitored. Federal regulations require evidence that each substance is safe at its intended concentration of use before it may be added to foods. Furthermore, all additives are subject to ongoing safety review as scientific understanding and methods of testing continue to improve. Consumers should feel safe about the foods they eat.

Figure 3.5: Processed meat having chemical preservatives.

http://s.eatthis-cdn.com/media/images/ext/359167724/processed-meat.jpg

NITRITES AND NITRATES

Nitrites give cured meats, like ham and bacon, the distinct color, aroma, and flavor we love. Nitrite is also an antioxidant and has antimicrobial properties, so they provide an essential food safety benefits, such as preventing the growth of Clostridium botulinum which causes botulism, one the deadliest foodborne illnesses. They are considered a preservative because their antimicrobial properties also control spoilage bacteria, so cured meats have a longer shelf-life than fresh meats.

Figure 3.6: Nitrites and nitrates in meat.

http://upl.stack.com/wp-content/uploads/2014/11/Sodium-Nitrate-Is-This-Controversial-Sandwich-Stuffer-Ingredient-Dangerous_STACK.jpg

Processors may choose to use nitrate or nitrite in their recipes. Nitrate (NO_3) is broken down by bacteria to nitrite (NO_2) during the curing process. Nitrates are only used in the long cures, for products such as dry sausages or dry-cured hams, which will continually need nitrate to produce nitrite throughout the curing process. For most of the cured meats, modern meat processors simply add nitrite directly to the formulation in very closely controlled amounts.

Natural Cure

Any processed meats labeled "natural" or "organic" cannot be prepared with nitrite because "preservatives" are not allowed in those products. As a result, there have been alternative curing processes that use natural sources of nitrate, such as celery juice or powder to create uncured products with a similar color and flavor to traditionally cured meats. The nitrate that is naturally within vegetables like beets and celery must be metabolized by specific types of bacteria and converted to nitrite. The nitrite then reacts with the meat to create the pink color and savory flavor characteristics associated with conventionally cured meats.

Confusion in the Labeling

Current USDA regulations require that meats cured with celery powder to be called "uncured" to distinguish them from conventionally cured products. Packages of meats cured with celery powder often say "No nitrates or nitrites added," but also contain a statement "other than those which naturally occur in celery powder." Some believe a more accurate way to define the products would be to call them something closer to naturally cured, but still must comply with the regulations as written, which requires them to be called "uncured."

Nitrites and Health

There is often a misunderstanding about the exposure to nitrites from meat. Of the body's exposure to nitrite, only 20 percent comes from food. Of that 20 percent, just 2 percent to 3 percent comes from processed meat. Nitrate is found naturally in green leafy vegetables such as spinach. Many vegetables are high in nitrate, which is then converted to nitrite when it comes into contact with saliva in the mouth. An evaluated 80 percent of nitrite exposure comes from that produced within the body.

CHEMICAL FOOD PRESERVATIVES

Preservation of provisions consisting of chemical foodstuff preservative is frequently stood to go on mutual otherwise numerous additives' synergistic movement, essential limitations for a product for example, composition, acidity, extrinsic factors e.g. processing heat, storage temperature, environment and water activity. Chemical foodstuff preservative is the resources that may be added in extremely small amounts almost 0.2% as well as that may not change organoleptic along with physical as well as the chemical characteristic of sustenance on otherwise simply extremely slight. While processing, or else expand through themselves throughout processes like chemical food preservatives, fermentation is added to foods like direct additives. Sodium chloride acids, common salt, sugar, alcohols in addition to the mechanism of smoldering are definite preservative have been utilized may intentionally otherwise accidentally used for centuries. In addition to conservation, these compounds contribute to the quality and identity of the products and are applied by processing procedures such as salting, curing, fermentation, and smoking.

This loom minimizes unwanted variations in manufactured goods characteristic and reduces the number of additives as well as the degree of processing managements. The perception of combinations of preservatives as well as treatments to preserve foods is frequently known as the barrier otherwise hurdle concept. Combinations of preservatives, as well as additives systems, supply limitless conservation substitute meant for functions in foodstuffs to assemble customer demands meant for harmless as well as healthy provisions. Conventional chemical food preservatives, as well as their utilization of vegetable and fruit processing technologies, could be reviewed as follows:

Ordinary saline: brined vegetables, Sugars like glucose, fructose, syrups, and sucrose: preserves, syrups, juice concentrates, jellies are foods conserved by means of high sugar concentrations.

Acidulants

Other preservatives and acidulates are produced in or added to fruit as well as vegetable products are as pursued:

Lactic Acid

It is produced by microbial degradation of sugars in foods like pickles and sauerkraut. Lactic acid is a chief item for consumption of a lot of foodstuff fermentation. Acidity created during the fermentation diminishes pH toward concentrations adverse in favor of an increase in the organism of spoilage like butyric-acid and putrefactive anaerobes producing bacteria. Sorbate and benzoate control the growth of these molds and yeasts.

Acetic Acid

The major application of vinegar, acetic acid includes products like sauces, ketchup, and pickles. It is the universal additive reducing a lot of yeasts, bacteria species as well as near molds of smaller degree. Even at identical pH amounts its preservative exploit is larger than so as to lactic acid and acetic acid is furthermore the lactic-acid fermentation product.

Other Acidulants

In some countries typically to acidify as well as preserve fruit sugar, jellies, jams etc malic as well as tartaric acids are used.

Citric Acid

It is a major tart present within citrus fruits physically and it's vastly utilized in effervescent beverages also like the acidifying mediator of food due to, its exclusive flavor characteristics. The other acids, it is a fewer efficient antimicrobial mediator. These have a limitless suitable everyday ingestion as well as are extremely water soluble.

Vitamin C otherwise ascorbic acid, its isomer erythorbic acid otherwise iso-ascorbic as well as in water their salts are extremely soluble in addition to secure in the direction of utilizing within foodstuff.

Universally used Food Preservatives

Sorbates

Sorbates are consumed as yeast, in addition to mold prevention in a diversity of food counting vegetables plus fruits, pickles, fruit juices, syrups, sauerkraut, jams, jellies, high moisture dehydrated fruits, preserves etc.

Benzoic Acid

One of the main general foodstuff chemical additions is inside the form of its sodium salt nature. In acidified otherwise acid provisions the same as syrups, fruit juices, jellies and jams, pickles, sauerkraut, fruit cocktails, preserves, and so on. Sodium benzoate is a universal preservative. By benzoate, yeasts are restricted to a better level than be bacteria in addition to molds.

Potassium Sorbate

While additional toward acidic food it's hydrolyzed in the direction of the acidic makes a fluffy powder, white and is extremely soluble within fluid above 50%. Calcium and sodium sorbates as well have stabilizer actions except their application is inadequate difference toward so as to use for potassium salt, which is in use due to its constancy, water solubility and common ease of preparation.

Sorbic Acid

It is normally measured non-toxic in addition to being metabolized; among other general food preservatives for sorbic acid WHO put the highest suitable daily ingestion per body weight 25 mg/kg. Salts of sorbic acid, as well as sorbic acid, are basically odorless as well as tasteless inside the food, while utilized on logical concentrations that are < 0.3 % and their antimicrobial activity are normally sufficient.

Gaseous Chemical Food Preservatives

Sulfur Dioxide Along With Sulphites

For many centuries, sulfur dioxide (SO_2) has been used the same as a fumigant and especially like a wine preservative. It is a suffocating, colorless, non-flammable gas, pungent-smelling as well as is in cold water 85 g in 100 ml that is extremely soluble at 25°C.

Sulfur dioxide, as well as a variety of sulfites, break up within water, as well as on small pH concentrations capitulate bisulfate, sulfite ions, and sulfurous acid. A variety of salts of sulfite consists of active sulfur dioxide 50-68%. The quantity of SO_2 ions raises with reducing values of pH and in water, a pH reliant balance is produced. The antimicrobial activity reaches its maximum at pH values less than 4.0. Within the variety of its bisulfite, sulfite, as well as salts of metabisulphite sulfur dioxide is used as a gas. By burning Sulphur otherwise through its release from the compressed liquefied form the gaseous form is produced. Metabisulphite is more firm to oxidation than bisulfites, which in turn show superior constancy than sulfites.

Sulfites, as well as sulfur dioxide, are utilized into the protection of foodstuff diversity. With several varieties being extra defiant than others, the antimicrobial exploit of sulfur dioxide in opposition to bacteria, yeasts, and molds is discerning.

(CO_2) Carbon dioxide

It is the colorless, odorless, and acidic in odor as well as flavor, noncombustible gas. During profitable observation, it's sold like a liquid in force 58 kg per cm^3 otherwise like dry ice solidified.

A source of low-temperature storage space as well as carrying of provisions, carbon dioxide is consumed like a dry, solid ice in many countries. Besides maintaining the low temperature, like it sublimes, the CO_2 gaseous reduce spoilage of the food like fruits and vegetables, etc as well as prevents psychrotrophic microorganisms growth.

In the storage of vegetables and fruits, it is utilized like an undeviating stabilizer. The correct combination of O_2 as well as CO_2 delays respiration as well as ripening retarding mold and yeast growth in the controlled modified environment storage of fruit plus vegetables. For the transportation as well as in favor of expenditure throughout the off-season, ultimate consequence is an extended storage of the yields. The amount of CO_2 (5-10%) is estimated by factors such as nature of product, variety, climate, and extent of storage.

Regulations for Chemical Preservation

Chemical food preservatives needed for a normal preservation also not more and have to be used just at a dosage level.

Chemical additives use ought to be there stringently restricted in the direction of the stuff that may be acknowledged like being devoid of unsafe sound effects happening on individual's health as well as be received through international as well as national legislation as well as standards. "Reconditioning" of chemically conserved foodstuff, for example, a new additives' accumulation so as to prevent the micro-biological decline previously found may be not suggested.

Factors Conclude the Act of Chemical Food Preservatives

Issue related to the chemical preservatives:

- Chemical composition
- Concentration

Micro-organisms related factors:

a) As a general decree it is likely to get the subsequent particulars like a basis for micro-organism species:
   - Sulfur dioxide in addition to its derivatives contain a sterile exploit on molds and on bacteria as well as yeast so could be taken as a "universal" preservative
   - Benzoic acid and its derivatives are stronger against bacteria than on yeasts as well as molds have a preservative exploit
   - Sorbic acid in higher dose levels acts on bacteria, except lactic in addition to acetic ones and also acts on molds and specific yeast species
   - Formic acid may be superfluous energetic in opposition to molds as well as yeasts along with fewer at the bacteria.

b) The efficiency of the chemical preservative is estimated in treated product by the initial number of micro-organisms.

The effectiveness is less if the product has been infected due to primary, not careful germ-free action or an incipient modification. So, the preservative dosage level could be reduced, through a little primary amount of microbes within the yield.

The product to be conserved, related specific factors:

- Effect of the pH intensity of the yield, effectiveness of the greater part of the chemical stabilizers may be elevated on lesser pH standards that are while the standard is extra tart.
- The chemical composition of the product.
- Physical presentation, size, dispersal via unit membrane taking place microbes as well as it evaluates the conservation consequence that the yield is serving towards the chemical additive's dispersal within foodstuff include an effect at the combination.

Therefore, the higher the preservative action when smaller is the slicing of the product. By gelatinous foods concentrated fruit juices etc preservative scattering is slowed down.

CHAPTER

4

NATURAL RESIDUALS IN CONTACT WITH FOOD

NATURAL CHEMICALS IN CONTACT WITH FOOD

Introduction

Food consists of a broad array of metallic elements, 'metals' for instance potassium, sodium, calcium, iron, zinc, boron, copper, selenium, in addition to magnesium. These substances are necessary for small amounts meant for protection of cellular tasks. Supplementary metallic essentials have no purposeful sound effects in the individuals as well as may also be damaging to physical condition, proviso foodstuffs including these are utilized frequently in the foodstuff. The preponderance of metals is normal components of the earth's coating.

Regarding damaging sound effects on health, the metals of particular apprehension are: lead (Pb), arsenic (As), tin (Sn), cadmium (Cd), as well as mercury (Hg), lead and mercury are frequently being referred to as "heavy metals" on account of their elevated atomic weight. Other potentially poisonous metals, for instance, are uranium as well as chromium have been also accounted to exist like contaminants in water otherwise food, as an amount of metals have been correlated with physical condition special effects in human beings uncovered to them in the place of work, for instance, nickel as well as beryllium.

HEAVY METALS

Toxicity of Heavy Metals

There are two major aspects regarding the toxicity of heavy metals:

(a) It is a fact that heavy metals have no known metabolic function, However, when they enter into a living body they disturb the normal cellular processes, hence, leading to toxicity of a number of body organs.

(b) The other major toxic aspect is the bioaccumulation, where some specific heavy metals, for example, mercury and lead, tend to accumulate in biological tissues.

Bioaccumulation due to the excessive exposure to mercury is related to a wide spectrum of adverse health effects to various body organs, including the damage to the central nervous system (neurotoxicity) and to the kidney.

The toxic effects of lead have been principally investigated during various studies conducted on a human who is exposed to lead during their work activities. Short-term exposure to high concentrations of lead can result in serious health conditions that may include brain damage, paralysis (lead palsy), anemia and gastrointestinal signs.

Cadmium has been associated with lung damage (including induction of lung tumors) and skeletal variation in occupationally exposed populations. However, the principal effect of **cadmium** is more associated to its toxicity to the kidneys.

Tin is relatively less toxic than mercury, cadmium, and lead. The major concern is the contamination of food being packed in the poorly manufactured tin cans, where tin can be leached into the food making it highly toxic to eat.

Arsenic exists both in inorganic and organic forms and also in various valence states. Inorganic arsenic is more toxic than organic arsenic compounds such as dimethylarsinate, and in turn, the trivalent forms of arsenic, e.g. arsenic trichloride, are more toxic as compared to the pentavalent arsenates.

Exposure to Heavy Metals in Diet

Metals such as mercury, cadmium, arsenic, and lead enter the environment primarily as a consequence of industrial emissions or through the disposal of products containing these metals, including mercury-cadmium or cadmium-nickel batteries, lead-containing ceramics and glass, mercury thermometers, etc.

Lead

As indicated, the probability of food contamination due to lead increases as a result of anthropogenic activities, such as mining and leaded petrol. However, like mercury, lead can cumulate in fish and shellfish and in addition can be found at higher levels in the offal (liver and kidney) of animals (used as food).

However, there are repeated examples of food dishes, utensils or other materials manufactured outside the EU that release lead into food at concentrations above the permissible levels allowed in the EU. Finally, the previous use of lead as a material for water pipes in many older houses may result in unacceptably high levels in water supplies. In, 1998 Drinking Water Directive and World Health Organization recommends and set a revised permissible concentration of 10μg/L for the lead in drinking water.

Mercury

The major potential dietary sources of exposure to methyl mercury are the sea foods (e.g., fish and shellfish in specific top predatory fish such as swordfish and marlin). The major cause for this high lead accumulation to marine organisms is the discharge of industrial effluents into the marine environment, followed by the uptake of lead by the microorganism. Once the lead is taken up by the microorganism, they convert the inorganic lead into more toxic organic methyl mercury. Methyl mercury has low breakdown rate and hence accumulates in marine organisms through the food chain, reaching potentially toxic concentrations in species at the top of the food chain, such as swordfish and marlin, which may then form part of the human diet.

Tin

Organotin compounds gain entry into a human body primarily through the diet (especially fish and fish products). Organic tin compounds are being used as antifouling agents and biocides in agricultural practices and once discharged into the aquatic environment becomes a part of the food chain as mentioned above for mercury exposure. Tin is Laos being used in cans manufacturing process to pack food and to protect the steel base from corrosion both externally (aerobic conditions) and internally when in contact with foods (anaerobic).

Arsenic

Various studies have been conducted to estimate the occurrence of arsenic in food. Recent studies on arsenic occurrence in the food explain that sea food accounts for over 90% of total exposure to arsenic in food. Further, with the exception of seafood and animal offal, the level of arsenic is generally less than 250 ug/kg. High levels are frequently found in crab, in which the white meat usually contains more arsenic than the brown meat.

Cadmium

Similar to the arsenic, cadmium also tends to accumulate in sea food (mussels, oysters and scallops etc.) and offal (kidney and liver) of mammals.

Analysis and Sampling for Mercury, Cadmium, Tin, and Lead in Food

Determination of trace metals involves two steps: Sample digestion and detection. Sample digestion may involve the following steps:

(1) Dry ashing in a conventional oven to eliminate the organic matter present in the sample

(2) Acid digestion to dissolve the metal contents of the sample (Acid digestion may involve Microwave digestion of the sample in a strong acid, dissolution of dry ashes directly in an acid by heating in a pressure vessel.

The detection methods used quite frequently are atomic absorption spectrometry (AAS) and inductively coupled plasma (ICP) techniques. Atomic absorption spectrometry includes flame AAS (CVAAS) for mercury determination and hydride generation (HG), generally used for arsenic determination. Inductively coupled plasma techniques commonly use mass spectrometry (ICP-MS) and ICP-optical emission spectrometry (ICP-OES).

NATURAL BIOLOGICAL/ BIOCHEMICAL RESIDUALS IN CONTACT WITH FOOD

Mycotoxins

Mycotoxins are natural toxins which are produced by fungi and may be toxic to humans and animals. They are formed by molds which grow on crops and foods under specific conditions. There are a number of mycotoxins present in the environment but only a few are present in foods and they are usually related to particular field crops like corn. The main prominent mycotoxins which cause health concerns in humans are aflatoxin, deoxynivalenol, ochratoxin, fumonisin, and patulin.

Some of the general characteristics of mycotoxins are following:

- Resistant to heat.
- Produced by fungi as secondary metabolites in response to competitive pressures from other fungi/bacteria.
- Can have antibiotic properties.
- Can cause toxic damage to cells of humans and animals.
- Can cause chronic effects such as various cancers, immunosuppressant, growth retardation, birth defects, and renal dysfunction.
- Can have serious long-term effects even in small amounts
- Usually related to certain crops (i.e. corn, cereal crops, apples).

Introduction to Mycotoxins

It is difficult to define mycotoxin in a few words. All mycotoxins are produced as secondary metabolites by filamentous fungi and are the natural products low-molecular-weight small molecules. While all mycotoxins are of fungal origin, not all toxic compounds produced by fungi are known as mycotoxins. Both the target and the concentration of the metabolite are very significant. Fungal products that are mainly toxic to bacteria (such as penicillin) are commonly called antibiotics. Fungal products that are toxic to plants are called phytotoxins by plant pathologists. Mycotoxins are made by fungi and are toxic to vertebrates and other animal groups in low levels. Other low-molecular-weight fungal metabolites such as ethanol that are toxic only in high levels are not considered mycotoxins.

More than three hundred different mycotoxins are known to date. However, the mycotoxins that are proven to be carcinogenic and/or toxic have gained specific scientific attentions. There are many pathways a human body gets exposure to mycotoxins, for examples by consuming plant-derived foods that are contaminated with toxins, animal products including meat and eggs that carry-over mycotoxins and their metabolites and exposure to air and dust containing toxins.

Cell biologists have classified the mycotoxins into different generic groups: teratogens, mutagens, carcinogens, and allergens. Organic chemists have attempted to classify them by their chemical structures (e.g., lactones, coumarins); biochemists according to their biosynthetic origins (polyketides, amino acid-derived, etc.); physicians by the illnesses they cause (e.g., St. Anthony's fire, Stachybotrys toxicosis), and mycologists by the fungi that produce them (e.g., Aspergillus toxins, Penicillium toxins). However, none of these classifications are entirely satisfactory.

Occurrence and Significance of Mycotoxins in Foods and Feeds

Aflatoxins (AFTs) are found in maize and peanuts, as well as in tree nuts and dried fruits. ATF is found mainly in cereals, however, significant levels of contamination may also occur in wine, coffee, spices and dried fruits. Other products of concern are beans, roasted coffee and cocoa, malt and beer, bread and bakery products, wines and grape juices, spices, poultry meat and kidneys, pig kidneys and pork sausages.

Aflatoxins

The aflatoxins were isolated and characterized after the death of more than 100,000 turkey poults. The investigation revealed the consumption of a mold-contaminated peanut meal. The major aflatoxins are called B_1, B_2, G_1, and G_2 based on their fluorescence under UV light blue or green and relative chromatographic mobility during thin-layer chromatography, M_1 and M_2 produced in milk and dairy products.

The liver is the primary target organ for Aflatoxins. Studies on poultry, fish, rodents, and nonhuman primates showed that when these were fed with aflatoxin B_1, it directly damages their liver. There are substantial differences in species susceptibility. Moreover, there are various factors (age, gender, weight, diet, exposure to infectious matter, the presence of other mycotoxins and pharmacologically active substances) that affect the response level of a given species to the aflatoxin toxicity.

Thousands of studies on aflatoxin toxicity have been conducted so far, and the major focus remains on the laboratory models or agriculturally important species.

Trichothecenes

The trichothecene mycotoxins (TCT) comprise a vast group of over 100 fungal metabolites with the same basic structure. Several fungal genera are capable of producing TCT; however, most of them have been isolated from Fusarium spp. All trichothecene contain an epoxide at the $C_{12,13}$ positions, which is responsible for their toxicological activity. At the cellular level, the main toxic effect of TCT mycotoxins appears to be a primary inhibition of protein synthesis. TCT effect actively dividing cells such as those lining the gastrointestinal tract, the skin, lymphoid and erythroid cells.

Moniliformin

Moniliformin i.e. a potassium or sodium salt of 1-hydroxy cyclo but-1-ene-3,4-dione is produced by several Fusarium species mainly F. proliferated and is usually found in the corn kernel. It can be transferred to next generation crops and survive for years in the soil.

Other Unidentified Mycotoxins

The impact of other mycotoxins on human health was reported in persons occupationally exposed to large amounts of different mycotoxin-producing fungi farmers, workers in silos, etc. In such cases, exposure to spores via the respiratory tract seems to be of considerable importance.

Toxoplasma gondii, infection from sheep or pigs, or mycotoxins was found in the grains. The proportion of late-term abortions gestational age 16±27 weeks was higher among farmers. The risk associated with grain farming was higher after the harvest, in seasons with a poor quality harvest and in pregnancies with multiple fetuses, which suggests that mycotoxins in grain induce labor at an early stage of pregnancy.

The patients were followed for periods of about one to ten years; they continued their work, avoiding massive reexposure to fungal dust, and during the observation period there were no further incidents.

Mycotoxins Negative Effects on Humans

Mycotoxicoses can be classified as acute or chronic like all toxic syndromes. Chronic toxicity is characterized by low-dose exposure over a long time period, resulting in cancers as well as other generally irreversible effects while acute toxicity generally has a rapid commencement and an obvious toxic response. Aflatoxicosis is poisonous hepatitis foremost to jaundice as well as, in rigorous cases, death. In Kenya for the duration of 1981, 2001, 2004 as well as 2005, India, in addition to Malaysia repetitive confrontation of this nature has occurred. AFB_1 have been comprehensively associated to human crucial liver cancer, in which it acts synergistically with HBV infection and was classified by the International Agency for Research on Cancer (IARC) as a human carcinogen. In developing countries, this combination shows a heavy cancer saddle.

Proceeding to the discovery and execution of modern milling carry out, Fusarium genus has been concerned in several human epidemics of mycotoxicoses. In alimentary toxic leukemia in Russia from 1932 to 1947 cereal grains unhygienic with *F. sporotrichosis* as well as *F. poae* were concerned. Indications included mucous membrane hyperemia, laryngitis, asphyxiation, vertigo, gastroenteritis, and oesophageal pain.

In China, Fumonisins also have been concerned in oesophageal malignancy. Though, with inadequate fundamental relationships and the existence of diverse impenetrable factors, data compiled by the International Agency for Research on Cancer were not decisive for F carcinogenicity in humans. Trichothecenes have been suggested as potential biological warfare agents. For illustration, the T-2 toxin was concerned like the chemical mediator of 'yellow rain' used in opposition to the Lao Peoples Democratic Republic from 1975 through 1981.

Experimental signs preceding death incorporated vomiting, blisters, hemorrhage, headache breathing difficulty, chest pain, diarrhea, fatigue, and dizziness. As well as nephritic congestion, autopsy findings integrated necrosis of the inside layer of the stomach as well as upper small intestine, liver, and lungs. However, it should be illustrious, that the cause of the samples of yellow rain is still a topic of dispute.

Factors Affecting Production, Contamination of Foods and Feeds, and Toxicity of Mycotoxins

A big and major difficulty which is faced in assessing the risk of mycotoxins to human and animal health is the multiplicity of factors affecting the production or presence of mycotoxins in foods or feeds. The presence of mycotoxins in foods or feeds can not be identified by just mere isolation and confirmation of mycotoxigenic fungal species

There are various physical factors (temperature, relative humidity, and insect infestation) that make the environment conducive for the colonization of fungal species and for the production of mycotoxin production. Chemical factors include the use of fungicides and/or fertilizers during agricultural activities that favor the production of mycotoxins.

The biological factors which are strongly based on the interactions between the colonizing toxigenic fungal species and substrate. While some plant species are more susceptible to colonization, environmental conditions may increase the vulnerability of other more resistant plant species.

In a three years study on the occurrence of aflatoxin in the selected USA foods (milled corn products, peanuts, and peanut products) and feeds (shelled corn, cottonseed, and cottonseed meal), the unpredictable nature of aflatoxin contamination and the difficulty in assessing the extent of such contamination have been documented. Very high levels (greater than the 20 μg/kg) of contamination were found in corn harvested from all parts of the USA in 1989, 1990, and 1991.

There are some anthropogenic factors such as chemical treatment and processing of food and feeds that may help to decrease the mycotoxin contamination. Aflatoxin in foods was found to be significantly decreased by using wet and dry milling processes and employing heat in the cooking process. Similarly, heating and roasting processes have been used to significantly decrease the aflatoxin contents in the corn. However, the literature review revealed that processing and pasteurization of milk do not completely destroy all the mycotoxins.

Recently, the potential role of dietary factors to counteract the toxic effects of mycotoxins has been reviewed. The role of antioxidants (Se and vitamins A, C, and E) and food additives was evaluated. Antioxidant defense mechanisms observed have included free radical scavenging, reduced lipid peroxidation, and general inhibition of the mutagenic process.

The antioxidant ethoxyquin has been recognized as a strong anti-aflatoxigenic agent. Scientists demonstrated the role of ethoxyquin in rat hepatocytes as induction of conjugating GST and also confirmed the enhancing effect of ethoxyquin on phase II metabolism in several subcellular components (microsomes, cytosol, and cell membrane) of the rat liver. In another study, gamma glutamyl transpeptidase was induced along with GST. A more recent study with marmosets has established ethoxyquin as a potential chemoprotective agent against the carcinogenic effects of AFB_1 in humans.

Economic Impact of Mycotoxins

Surveillance studies showed that worldwide contamination of cereal grains and other feeds with Fusarium mycotoxins is a global concern. In Yugoslavia, studies on mycotoxigenic fungi in raw milk have indicated that 91% of the samples tested were contaminated. In the USA, a study was conducted in seven Midwestern states in 1988–1989 and found mycotoxins in 19.5% of corn samples assayed prior to any induced environmental stress and 24.7% of the samples following stress induction. It was estimated the 1980 losses due to AF in corn of eight Southeastern states at 97 million dollars with additional 100 million dollars in production losses at hog farms feeding the contaminated corn.

It was estimated that 10 million dollars were lost in India's export within a decade due to groundnut contamination with mycotoxins.

Guidelines of Mycotoxins in Feed and Foodstuff

Many developing countries have realized that reducing mycotoxins concentrations in foods will not just decrease the financial burden on health care but also grant international trade recompense like exports to the striking European markets. Other than the direct economic losses, health risk and suggestions arising from mycotoxicoses are huge. Factors fundamental to country's ability to protect its population from mycotoxins include the capability to test food for contamination as well as the supporting will to address mycotoxins contact which determines whether requirements can be imposed.

Good Agricultural Practices

Agronomic practices have shown to include the deep effect on mycotoxins contagion of crops in the pasture.

Early Harvesting

Early harvesting minimizes fungal illness of crops in the field before yield and resulting contagion of harvested manufacture. Farmers must have authentic and to necessary information about harvesting time and patterns. Yet the majority of farmers in Africa are well aware of the need for early harvesting, need for cash, labor constraint, unpredictable weather, the threat of thieves, rodents as well as other animals force farmers to harvest at an unsuitable time.

Proper Drying

Fast drying of agricultural yields too low humidity concentration is dangerous as it creates less encouraging situations for fungal growth as well as proliferation, insect infestation, and assets remain longer. It was reported that drying harvested maize to 15.5% moisture substance otherwise lower contained by 24–48 h would minimize the risk of fungal growth as well as resulting aflatoxin manufacture.

Physical Treatment

Proper physical treatment is also very important. A study which was conducted in Benin in order to evaluate the fortune of aflatoxins Furthermore fumonisins by traditional processing of naturally-contagin maize and maize based foods demonstrated that sorting, washing, winnowing, crushing joint with de-hulling of maize grains were effective in achieving essential mycotoxins elimination. Study of the allocation of aflatoxin in peanuts shows that a chief portion (80%) of the toxin is mostly related with the small in addition to shriveled seed as well as moldy.

Sanitation

Basic sanitation measures like removal and demolition of remains from preceding harvest would assist in minimizing illness as well as invasion of produce in the field. Cleaning stores before loading new product have been shown to be associated with fewer aflatoxin levels.

Proper Storage

To protect assessment in storage, it is essential to prevent biological activity through adequate drying to less than 10% moisture, inert atmospheres, eradication of insect movement that can raise moisture content through condensation of moisture resulting from respiration and low temperatures.

Insect Management

Insects' bear spores of mycotoxins posing fungi from plant surfaces to the center of the stalk otherwise seed or produce infection wounds by their feeding behavior. The limit of insect harm influences the degree of mycotoxins contagion. It was found that insect hurt of maize is a good forecaster of Fusarium mycotoxins contagion. Consequently, proper organization of insect vermin through any suitable control policy would minimize mycotoxins infectivity dilemma.

Decontamination

Decontamination is attributed to the removal of contamination by adopting proper techniques and protocols. Chemoprotection or Anterosorption could be the options to decontaminate the food or feed contaminated with mycotoxins. Chemoprotection of aflatoxins is carried out by using a number of chemical compounds like oltipraz and chlorophyllin or dietary intervention like broccoli sprouts and green tea that either increase an animal's detoxification processes or prevent the production of the epoxide that leads to chromosomal damage. Entersorption process for the decontamination of food and feed is based on the discovery of certain clay minerals, such as Novasil, which can selectively adsorb mycotoxins tightly enough to prevent their absorption from the gastrointestinal tract.

Breeding for Resistance

Potential biochemical and genetic resistance markers have been identified in crops, particularly maize in different parts of the world. These aflatoxin resistant markers are being utilized as selectable markers in breeding to avoid contamination.

Prototypes of genetically engineered crops have been developed with the purpose:

(a) to reduce the fungal virulence and to reduce the phytotoxic effects of certain trichothecenes, these prototypes contain genes for resistance

(b) for reducing fungal infection in the USA, these prototypes contain genes encoding fungal growth inhibitors. Gene clusters housing the genes that govern the formation of trichothecenes, fumonisins and aflatoxins have been elucidated and are being targeted in strategies to interrupt the biosynthesis of these mycotoxins.

Conclusion

Illness is being caused by fungi in various ways. Mycoses are the best-known diseases of fungal etiology, but toxic secondary metabolites produced by saprophytic species are also a significant human health hazard. The term mycotoxin is an artificial rubric used to describe pharmacologically active mold metabolites characterized by vertebrate toxicity. They fall into various chemically unrelated classes, are produced in a strain-specific way, and elicit some complicated and overlapping toxigenic activities in sensitive species that include carcinogenicity, inhibition of protein synthesis, immunosuppression, dermal irritation, and other metabolic perturbations. Mycotoxins usually enter the body through different ways such as ingestion of contaminated foods, inhalation of toxigenic spores and direct dermal contact.

In conclusion, mycotoxicoses will remain diagnostically daunting diseases due to inappropriate investigation strategy and non-reliable laboratory tests.

Glycoalkaloids

Potatoes can contain natural toxins called glycoalkaloids. The major ones present in potatoes are α-solanine and α-chaconine. These toxins are produced in response to stresses such as UV light and damage (such as bruising), and cannot be destroyed by cooking. Toxin levels are highest in the peel and sprout of the potatoes and can be seen as a characteristic green color on those components.

Exposure to glycoalkaloids can cause critical toxic effects such as burning in the mouth, diarrhea, severe stomach ache, vomiting, and gastrointestinal irritation. Death from glycoalkaloid poisoning is rare.

Other Natural Toxins

Other natural toxins are listed in the Table below.

Table 4.1: Some Natural Toxins in Food Plants

Food commodity	Toxin
Ackee fruit	Hypoglycin
Cassa root, bamboo shoots, stone fruit	cyanogenic glycoside
Fiddlehead	Unidentified
Green beans, red kidney beans, white kidney beans	Lectin
Wild mushrooms	amanitins, gyromitrin, muscarine, phallotoxins
Parsnip	Furocoumarins
Rhubarb	oxalic acid

Cabbage, cauliflower, broccoli, mustard, turnip	Goitrogens

Marine Toxins

Marine toxins are a category of toxins that sometimes cumulate in fish and shellfish. There are two origins of marine toxins:

- Decomposition
- Microscopic marine algae (phytoplankton, including diatoms and dinoflagellates)

Decomposition

When certain fish, especially scombroid fish (i.e. tuna, bonito, and mackerel), start to decompose, histamine is formed. Histidine, a naturally present amino acid, is changed into histamine by an enzyme produced by specific bacteria during decomposition. Histamine, in small doses, is essential for the proper functioning of the human immune system. However, histamine in higher dose may trigger severe reactions when consumed same to those seen in allergic reactions such as rash, nausea, vomiting, diarrhea, headache, dizziness, burning throat, stomach pain and itchy skin. The presence of high amounts of histamine indicates that decomposition has occurred, even if the decomposition is not clear. Toxic levels of histamine can form before a fish smells or tastes bad. As this species only occurs in scombroid fish, this is called Scombroid Poisoning.

Microscopic Marine Algae

Many marine toxins are produced by and can cumulate in fish and shellfish if they ingest certain kinds of algae.

Table 4.2 lists some common marine toxins and the seafood they are commonly associated with.

Table 4.2: Common Marine Toxins

Illness	Toxin	Seafood
Paralytic shellfish poisoning (PSP)	Saxitoxin	Oysters, clams, scallops, mussels, cockles, whelks

Amnesic shellfish poisoning (ASP)	Domoic acid	Bivalve molluscan shellfish, clams, mussels, oysters, scallops
Ciguatera poisoning	Ciguatoxin	Tropical fish such as barracuda, amberjack, red snapper, grouper
Diarrhetic shellfish poisoning (DSP)	Okadaic acid	Various shellfish, cockles, mussels, oysters
Tetrodotoxin poisoning	Tetrodotoxin	Pufferfish, California newt, parrot fish, octopus, starfish, angelfish, and xanthid crabs
Neurotoxic Shellfish Poisoning	Brevetoxin	oysters, clams, and mussels

Sodium Nitrite and Nitrosamines

In the manufacturing of processed meat products, sodium nitrite substance is used extensively. As a colorant and preservative, it is added. It keeps meat that would otherwise turn brown or gray, a more pleasing shade of red or pink keeps it for the way food manufacturers particularly like.

Figure 4.1: Sodium nitrate in meat.

http://www.healthline.com/hlcmsresource/images/topic_centers/Food-Nutrition/642x361-Is_Sodium_Nitrate_Bad_for_You.jpg

Though, there's a big problem with sodium nitrite is while use in products of processed meat. It has been shown to mingle with a functional group so as to exist in proteins recognized as amines, at high temperatures as well as make toxic substances identified like nitrosamines. To boost the danger of colon, stomach and pancreatic tumors, these nitrosamines are believed carcinogenic compounds.

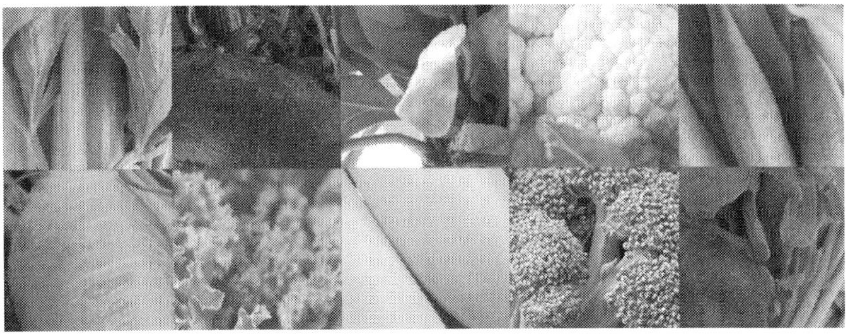

Figure 4.2: Nitrites in vegetables.

http://www.dole.com/~/media/ArticleImages/Nitrates-Nitrites-help-or-harm_1338x460.jpg

In several vegetables may as well be altered to nitrites in the individual stomach, occasionally protectors of sodium nitrite will point to the detail so as to obviously taking place nitrates. In processed meat, they then use this to create the case so as to sodium nitrite ought to be safe.

However, vegetables aren't typically heated to the identical type of warmth as well as don't have the amines of meat products. Due to this, they are a lot fewer possible to guide to the source of carcinogenic nitrosamines. Actually, by reason of the antioxidant material similar to vitamin C and beta-carotene in various vegetables, they are essentially more likely to supply a little security in opposition to the sodium nitrite in processed meat.

Can Vitamin C Make Less Harmful Processed Meats?

Learning has exposed to vitamin C might supply various protections as of the development of hazardous nitrosamines, as the finest suggestion is to slice down on your processed meat expenditure.

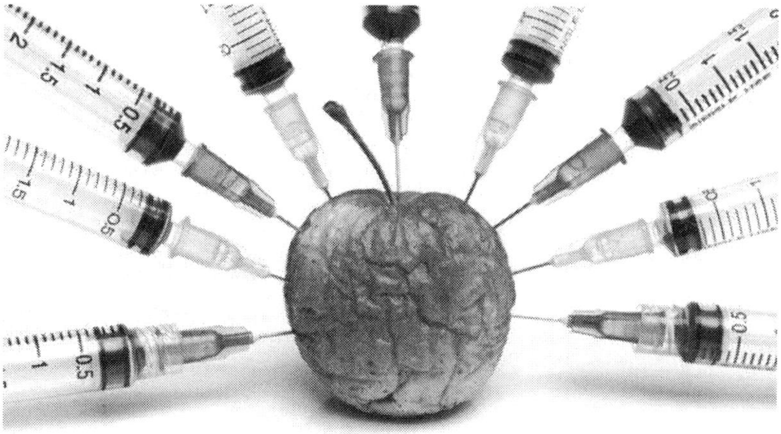

Figure 4.3: Additives in fruit.

https://www.google.com.pk/url?sa=i&rct=j&q=&esrc=s&source=images&cd=&cad=rja&uact=8&ved=0ahUKEwj26KC97_PUAhWBuhQKHU5lAYIQjRwIBw&url=http%3A%2F%2Fwww.grandparents.com%2Fhealth-and-wellbeing%2Fdiet-and-nutrition%2Ftoxins-in-food&psig=AFQjCNGPnNaOOLHt2b5u4CiFOxhleAZgcQ&ust=1499403285887164

Researchers have found that when vitamin C is in use at the similar time like a food with processed meat, as bacon or pizza, it can limit nitrosamine production in stomach and also is a powerful antioxidant.

Definitely no support to consume processed meat used for breakfast, lunch as well as dinner, and this is not measured totally valuable. Though, a good quality vitamin C enhancement may assist to level any possible harm if you are going to contain a meal with sodium nitrite in it.

Ascorbic acid isolated as of GMO corn is not the mainly valuable sort of vitamin C that is present in the majority of multivitamins. Vitamin C must be in its usual form, for finest incorporation.

The mainly useful accepted variety of vitamin C is greatly regarded like tropical camu camu extract, the food have processed meat, is the finest to use with. It's a superior scheme to remain a number of these capsules at work, otherwise even in a purse or carry bag used for while you're ingesting out, because several hours afterwards will be too late.

Figure 4.4: Additives in processed meat.

http://s.eatthis-cdn.com/media/images/ext/359167724/processed-meat.jpg

Alternatives of Processed Meat

The grass fed steak; wild salmon, organic chicken and other fit meats are not affected by the sodium nitrite, the difficulty just associated to processed meats from pork otherwise beef. They are as well a lot of nourishing for assortments of further causes moreover choose to have meats in the diet these of the superlative type to be eating. In the intervening time, the ginger, garlic as well as chili marinated chicken forms a tasty substitute in favor of strongly tasty processed meats, without damaging potential nitrosamines from sodium nitrite otherwise sodium chloride.

Figure 4.5: Meat.

http://www.finermeatco.com/uploads/9/1/3/9/9139697/7450703_orig.jpg

Cutting Back on Processed Meats

It's best to decrease on processed meats for long-term fitness, processed meat, the origin of carcinogenic nitrosamines among a fine natural vitamin C.

Insert to the elevated levels of destructive sodium nitrite, sodium chloride, nitrosamines in addition to the tumor hazard associated to consumption of processed meats.

Food Allergens

By combating off contagions as well as additional threats to good fitness, the immune system of body maintains health. When the immune system exaggerates to a foodstuff otherwise any material within a foodstuff, recognizing it like a hazard as well as prompts a shielding reaction, a foodstuff allergic response takes place.

As it is impossible to forecast whether a youngster will take over a foodstuff allergy of parents otherwise whether siblings will have a situation, stipulation allergies able to run in family units. Several researches do recommend so as to a child's younger siblings with a peanut sensitivity will as well be sensitive to peanuts.

Symptoms of a foodstuff allergic reaction may sort as of gentle to harsh. Just as a preliminary response causes little disruptions doesn't represent so as to the entire reactions will be alike; a foodstuff that activated simply gentle indications on an event might pose extra serious indications at a different instance.

The mainly adverse sensitive to reaction is 'anaphylaxis' a serious whole-body sensitive to response so as to be able to damage the breathing, pose a remarkable fall to the blood pressure as well as influence the heart speed. Anaphylaxis may approach inside moments of contact to the activate foodstuff. It may be lethal as well as ought to be treating rapidly by an inoculation of epinephrine, adrenaline.

As any foodstuff may pose an unfavorable effect, eight types of foodstuff in favor of almost 90 percent of the responses:

- Milk
- Eggs
- Tree nuts

- Peanuts
- Shellfish
- Fish
- Soy
- Wheat

Some kernels, counting mustard seeds, the main element in the condiment mustard as well as sesame too are common foodstuff allergic reaction promptors in addition to measured the chief allergen in several states.

Symptoms of the sensitive reaction might have the gastrointestinal tract, the skin, the respiratory tract as well as the cardiovascular system. These may takes place in one otherwise supplementary of the subsequent customs:

- Hives
- Circulatory collapse or Shock
- Stomach cramps and/or vomiting
- Wheezing
- Breath shortness
- Hoarse, tight throat; difficult swallowing
- Repetitive cough
- Weak pulse
- Swelling of the tongue, affecting the ability to breathe otherwise talk
- Feeling faint or dizziness
- Blue or pale coloring of skin
- Anaphylaxis, a potentially acute response so as to may damage inhalation as well as propel the organization into distress; response might concurrently influence different components of the body for instance, a indigestion convoyed by means of an itchiness.

Mostly foodborne illnesses indications take place in two hours of intake; mostly they initiate within short times. The response might be overdue by means of four to six hours otherwise even longer, in several exceedingly exceptional cases. In children, belated responses are most repeatedly seen who build up eczema like the symptom of foodstuff allergic reaction as well as in people by an unusual allergic reaction to red animal protein caused through the taste of a single star mark. An additional sort of late foodstuff sensitivity response stalks as of food protein-induced enter colitis syndrome (FPIES), a serious gastrointestinal response so as to in general happens two to six hours after overwhelming soy, milk, some particles as well as various other solid foodstuff. It typically establishes in little newborns that are being uncovered to this foodstuff intended for the earliest moment otherwise who are being weaned. FPIES may lead to lack of fluids as well as frequently occupies recurring nausea. Babies will have bleeding in diarrhea, in a few cases. As the signals be similar to those of a bacterial illness otherwise viral ailment, analysis of FPIES might be late. FPIES is the medicinal urgent situation so as to must be treated with rehydration.

Not everybody who got indications after ingestion of precise foodstuff has a foodstuff allergic reaction otherwise require avoiding so as to victuals completely; for illustration, several natives incident an uncomfortable oral cavity as well as esophagus after consumption of an uncooked otherwise raw vegetable or raw fruit. This might notice oral sensitivity syndrome, a response to pollen, not to the certain foodstuff itself. The immune system recognizes the pollen as well as same proteins in the foodstuff moreover expresses an allergic sensitivity to retort to it. The allergen is damaged by means of appliying warmth to the foodstuff that may then be used with no disruption.

Salmonellosis

Salmonellosis is a type of food poisoning caused by the Salmonella enterica bacterium. There are many different types of these bacteria. Salmonellaserotype Typhimurium and Salmonella serotype Enteritidis are the most general kinds in the United States.

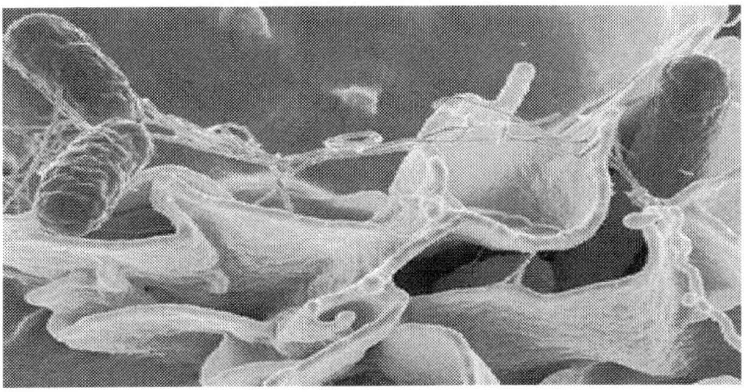

Figure 4.6: Salmonella.

http://img.webmd.com/dtmcms/live/webmd/consumer_assets/site_images/articles/health_tools/salmonella_outbreak_tomatoes_slideshow/PRinc_rm_photo_of_salmonella1.jpg?resize=300px:&output-quality=100*

Salmonellosis is mostly common in the summer than in the winter. Children are the most likely to get salmonellosis. Young children, older adults, and people who have impaired immune systems are the most likely to have critical infections.

Infection Causes, Symptoms, Treatment, and Prevention of Salmonellosis

You can get salmonellosis by eating food contaminated with salmonella. This may happen in the following ways:

- Food may be contaminated while food processing or food handling.
- Food may be contaminated by the unwashed hands of an infected food handler. A frequent pose is a food handler who does not wash his or her hands with soap after using the bathroom.
- Salmonella may also be present in the feces of some pets, especially those with diarrhea. You can become infected if you do not wash your hands after contact with these feces.

Beef, poultry, milk, and eggs are most often infected with salmonella. But vegetables may also be contaminated. Contaminated foods usually look and smell normal.

SUMMARY

Metals as well as supplementary essentials are able to be usually present in foodstuff otherwise are able to come into foodstuff as the consequence of human actions for instance agricultural as well as industrial developments. The metals of appropriate apprehension regarding dangerous sound effects on fitness are lead, mercury, tin, arsenic in addition to cadmium. Lead as well as mercury is frequently referred to the same as "heavy metals". The toxicity of these metals is in fraction due to the reality so as to they accumulate in natural tissues, a method calleed **'bioaccumulation'**. This method of bioaccumulation of metals happens in the entire living individuals as a result of introduction to metals in foodstuff as well as the surroundings, counting provisions natural world like cattle as well as fish and in humans.

So as to make sure that these MLs may not surpass custom observation of foodstuff must be approved, concerning the captivating of illustration of potentially impure manufacture, pursued by laboratory examination to evaluate the concentrations of the metal in query in the manufactured goods. Various other toxins may as well infect the foodstuff at any juncture of the foodstuff manufacture, handing out as well as covering as well as be able to pose a lot of persistent, long term otherwise short term ailments when they are consumed over than the definite specified quantity. A small amount of these can pose much damage yet at little amounts. The producers as well as consumers ought to be having sufficient data concerning these levels, in order that to get rid of the menace of ailments.

CHAPTER

5

EVALUATION OF SENSORY PROPERTIES OF FOOD

Smell, sight, taste, kinesthesis, pain, touch and temperature are the senses by which we perceive food. These senses are often not noticed or confused. When we bite into food, we are mostly unaware of the differences among the feel, taste, and odor ambiances. The sensation received from the eating a peach is little more than the sweetness, a slight bitterness, a slightly sour taste, and the cooling ambiance of the pulp if the nose is tightly closed. Mainly taste and touch are what we generally describe to odor.

Sensory estimation does not only compact with "likes and dislikes," "OK or not OK" however the development systematically measures, extracts investigate as well as deduces the physiological as well as psychological answers to substantial spur formed as a result of a foodstuff.

Sensory estimation is a systematic regulation those investigate as well as dealings individual reactions to the nature and constituents of drink as well as foods.

BRIEF INTRODUCTION

'A systematic regulation employed to measure, remind, examine as well as understand responses to the characters of foodstuff as well as substances like they are supposed as a result of senses of touch, smell, sight, taste, and hearing.'

Our awareness of the sensory psychology (and of sensory physiology) is still just in the developmental stages. Food scientists will be capable to form more perceptive measurements of human response to foods when enough essential information available. This becomes of critical importance when we intake, apparently, a duration of lack of energy, as well as safe, new, nonanimal sources of sustenance have to be developed. Food scientists must ensure that new sustenance is not only nutritious but also be appealing to the consumer as regards color, taste, smell, flavor, appearance, and feel.

'Sense' may be expressed as the physiological observation of a stimulus.

In human beings, there are five senses:

Sight: Capability of the brain as well as eye to sense electromagnetic impressions inside the noticeable collection of radiance as well as construe the illustration.

Hearing: Sound sense. When sensations transmit all the way all the way through a channel (for example, atmosphere) is perceived by way of the mind, echo is supposed.

Contact: Mostly in the tongue/skin, sense of pressure perception.

Taste

Smell

SENSORY FACTORS

Gustation / Taste

It is a "chemical" sense. In the tongue, the receptors (buds) might differentiate five flavors. Salt, bitter sweet, sour and umami that is a delicate and savory flavor so as to be related to a brothy or soapy note. The receptors on the tongue recognize the glutamic acid remains in the foodstuff.

Taste System

The main taste receptor is the tongue.

The organizations so as to offer the tongue its irregular organization through elevated projections on the exterior are known as papillae. In the human tongue, four sorts of papillae found:

- Fungiform papillae
- Filiform papillae
- Foliate papillae
- Circumvallate papillae

Except for the filiform papilla, taste buds are present in all of the papillae. The tongue has around 10,000 taste buds.

Taste System Tongue Surface with Papillae

Every taste bud has taste cells so as to enlarge the taste minute opening. Each taste cell is connected through one otherwise other nerve fiber. Ion associations diagonally the covering of flavor cells occur to grounds transduction when chemicals as of foodstuff get in touch with the pours of taste cells. By a lot of passageways, these signs arrive at the spinal cord as well as after that set off to the Thalamus. Then, the mind notice the taste of the foodstuff

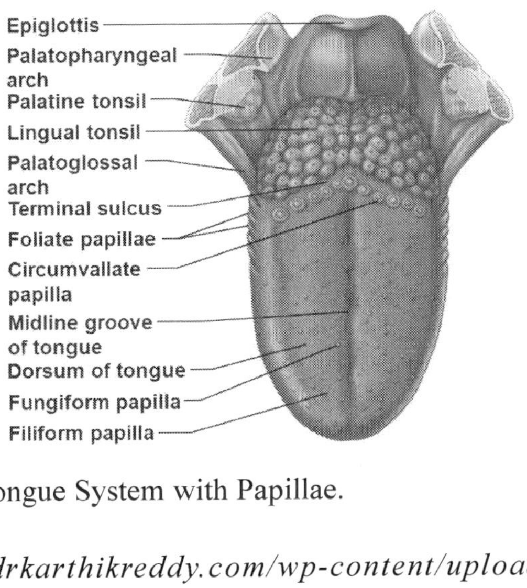

Figure 5.1: Tongue System with Papillae.

http://www.drkarthikreddy.com/wp-content/uploads/2016/03/tongue-e1458049185343.jpg

Taste is articulated in terms of provisions being sweet; umami, weak, cool, metallic, bitter, spicy, zesty, savory, warm, mild, hot, citrus, tangy, strong, sour, acidic, tart, sharp, rancid, rich, bland and salty.

Smell / Odor

Volatile aromas released from food are detected by the nose. A definite aroma might be illustrated in favor of a fastidious foodstuff, for example, nutty, cheesy and green and so on. The strength might as well be traced. Odors, as well as the flavor of a food, together make the taste of a foodstuff. Because of these means, public suffering as of cold gets it hard to estimate the tastes in foodstuff.

Generally, odors are illustrated while fragrant, aromatic, mossy, pungent, scented, musty, spicy, weak, woody, buttery, floral, mild, bland, strong, green, acrid, citrus like oily, creamy, earthy, tart, rancid, rotten, leathery and savory.

Smell Discipline: Olfaction

Individuals are macrosmatic, humans do not possess an enthusiastic common sense of aroma like it is not vital in favor of our continued existence. Humans are able to distinguish amongst 100,000 aromas however they are unable to tag them precisely. Primary, aroma molecules as of foodstuff connect to receptors in the nose. Indications as of the receptors pass through the olfactory bulb, a Q-tip-like organization generally over the eyes.

As of there, several signs go off to the chief olfactory cortex as well as on to the higher-order components of the mind. Except for a few signs as of the olfactory bulb straightforwardly set off to the amygdala in the mind, a region so as to be applicable to feelings as well as an amount of go to the hippocampus that is concerned with remembrance. Therefore, we occasionally connect a few aromas in the recall with an explicit feeling. Acknowledgment of aroma is instant like the olfactory reaction is instantaneous, widening unswervingly in the direction of the mind.

It is the single consign wherever the middle nervous organization is openly uncovered in the direction of the surroundings.

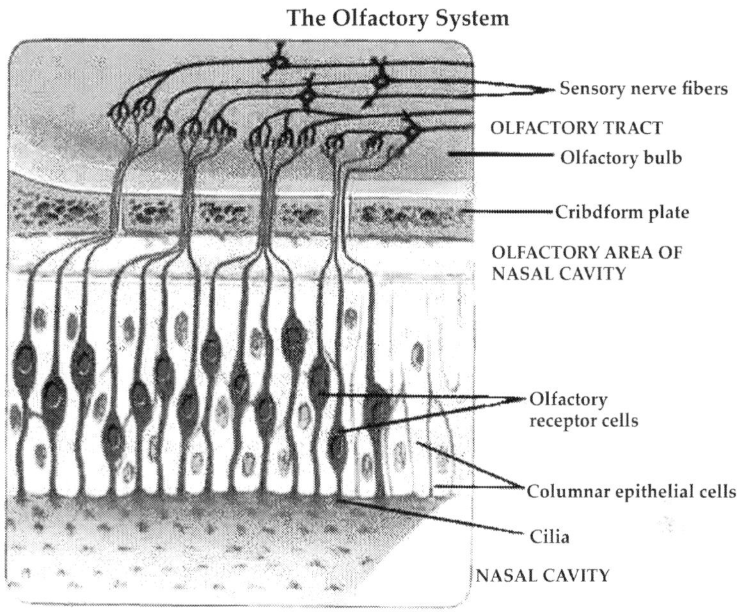

Figure 5.2: The Olfactory System.

https://thegrilledatheist.files.wordpress.com/2012/08/smell.jpg

Sight/Appearance and Color

In foods, color is particularly significant; the consumer may never taste otherwise touch the food as if the color is unattractive. Against dirt on food was one of the earliest discriminations. In the beginning, it might have been based on the avoidance on unpleasant contagious like sand grains as well as on the sense of touch, but this prejudice is now largely visual. In 1944, Renner also points out that the dirt color is generally dark, and these colors are commonly considered unpleasant, probably due to the growth of mold is mostly dark, as in fecal matter.

***Color Discrimination*:** Often with the valid basis, color prejudice among the individuals exist, Red, as well as yellow, are associated with desirable flavor as well as ripeness, green fruit is commonly unripe. A specific color (or lack of it) is associated not only with the processed quality of the food but also with the native quality. In blue cheese, lack of adequate mold, meat mold spoilage, in white table wines, an amber color, in canned corns, brownies all reveal the negative color-quality factors. The purple heme pigment myoglobin is the pigment responsible for the color of fresh meat. On oxygen exposure, it forms bright red oxymyoglobin. A grayish brown metmyoglobin is an oxidized pigment. In the meat department display, use of the fluorescent fixtures counters shifts the visible color of the meat towards red, making it appear less oxidized and thus more appealing.

The only quality factor or the most significant factor the consumer recognizes is the proper color. As an important measure of quality appearance as well as the color is widely accepted for the alcoholic beverages. In fruits, vegetables, as well as fish the appearance and color, are often good quality indicators specifically their maturity otherwise freshness. Off-color wines may be too old or too young or may have been improperly processed as well as cloudy wines are commonly spoiled. Some markets traditionally paid more for brown eggs than for white, but here the basis of the quality-color estimation is not known.

Because of aesthetic considerations, the color of food also may appeal. The decoration on cakes, the concurrence of colors on the plate as well as the combination of colors salads all contribute to our appreciation of foods. We classify beer or wine by color, we judge the strength or weakness of tea or coffee by the color. We insist on color in some foods that we express disapproval of in others: green egg white would be nauseating but blue and green plums and blue cheese are appreciated and accepted.

The U.S. Food and Drug Administration (FDA) considers any added color, artificial or natural, to be a food additive. It must be noted on the label as well as must be specifically approved for use. The labeling required depends upon whether the color is artificial otherwise natural. Any measurable amount that may have been added must be noted. Where for a good standard of identity has been developed, if that standard does not contain a statement of added color, the color additives may not be used.

Our impressions of the physical world recorded by eyes are size, color, gloss, and shape, glitter and so on. Though the eye is not a very good quantitative instrument, however, it is very good qualitatively.

Color Measurement: By a single parameter, color cannot be expressed. We must determine dominant wavelength, colorimetric purity, and intensity, to define a color. Roughly, the dominant wavelength is the predominant color (e.g. red, yellow), the colorimetric purity is the relative amount of gray present, and intensity is the amount of the color. Either the light transmitted through the food or reflected from the surface is analyzed. For appearance, there are two categories: geometric attributes related to the spatial distribution of light, and color attributes related to wavelength distribution.

The eye has its physiological attributes as well as its limitations, like the other sense organs. The most significant of these properties are 1) differential threshold 2) adaptation 3) absolute threshold 4) duality of reception 5) hue and saturation discrimination. At the threshold, the effective light energy is 5-11 quanta (average 7) or a few hundred billionths of an erg (unit of energy). This means that the eye is so sensitive that if it were much more so, we could 'see' the shot photon emission, and 'steady' light would not be steady.

Color maps are used to provide reference points and various systems of color nomenclature are available. The 'color dictionary' of Paul and Maerz contains 7000 color samples under 4000 names. In the United States, the Munsell system is widely used. Finally, the tristimulus system of the International Commission on Illumination (ICI) is generally accepted by food scientists. Brightness (amount of color) and purity (saturation) are obtained from an absorption curve hue (dominant wavelength).

For many foods color standards are present that are detected by eye, the sight sense.

Sound

To food scientists, the sound is not of great importance but it must not be dismissed as insignificant. The importance of the sound can be determined by the crunching of some foods like the crunching of nuts is a significant sensory impression. The crunching of the nuts or the other food products like potato chips distracts attention by the sound and feel of taste because of the sound. Most probably, the crunchy potato chips are crisper.

Touch

Mostly, the sense of touch is defined as the factor in the appreciation of foodstuff. However, consider the graininess of pears or nuts, the sogginess of wet bread, the importance of the creaminess of the chocolate and the crumbliness of cake. Touch assists to recognize our food as the safety of the food is concerned but it assists us to reject bulky or sharp provisions.

Conclusion

Indeed, sensory appraisal of food is an extremely definite as well as multifarious knowledge. It has to be memorized so as to no section of laboratory tools might contest the capability of individual sanity of touch, odor, as well as taste along with a prospect to determine the slight disparities within foodstuff texture, taste, appearance as well as flavors. It is suggested so as to the sensory estimation constantly goes hand in hand by means of influential investigation in foodstuff developing corporations as well as laboratories.

PSYCHOLOGICAL FACTORS

These factors recount in the direction of the brain as well as the sensations. They are complicated to explain, as well as vary as of individual to individual conditional on their way of life in addition to education. Various psychological factors like habits, values, beliefs as well as past understandings by means of foodstuff encompass a continuous persuade on the foodstuff chosen, as options prepared the same as the consequence of self-concept, attitudes as well as emotions might be different as of every day.

Values

The value is a profound individual emotion concerning what is significant. Values are powerful as much as necessary to manipulate actions along with encouraging actions. The person's morals might be a sign of those of the relations as well as customs in which they were increased; otherwise, they might be the individual's reaction to the understandings come across all through the life. In conditions of foodstuff collection, the standards mainly like to persuade selections are associated in the direction of foodstuff beginning as well as the upholding of physical condition.

They revere the privileges of all livings to live in serenity, free as of pain. Vegetarianism is habitually an indication of value found foodstuff choice. An individual might discover the consideration of assassination as well as consumption of a living thing disgusting; otherwise, they might diverge by means of the circumstances under which several animals are increased the same as a foodstuff supply. Several inhabitants' value an animal's right to liberty as a result extremely so as to they do not have the same opinion through the utilization of some living thing foodstuffs, counting milk, eggs, otherwise additional dairy foodstuffs.

Beliefs

Beliefs concerning what is satisfactory to consume differ all through the human race as well as are frequently associated on the way to religious conviction as well as civilizing inheritance. A belief is a confidence otherwise estimation which must not be based on optimistic systematic evidence. Beliefs might be changed as well as challenged, not like standards so as to stay unchanging. Most of the religious convictions contain foodstuff civilizations as well as impose limitations on what their groups consume. For instance, Buddhists are incomplete vegetarians. They might consume eggs; fish, as well as dairy foodstuffs, however, are not permissible red meat. (They consider it is mistaken to say otherwise harm living possessions additional than fish.)

Figure 5.3: Psycological Belives Affecting Sensory Properties of Food.

http://factorsaffectingfoodselection.weebly.com/uploads/1/8/3/7/18371025/9980566.jpeg?282

Over and above spiritual traditions, several communal assemblies, as well as societies, have definite attitudes regarding foodstuff. Many of these attitudes cannot be authenticated as well as contain small if several, dietetic basis; a lot of the faddist. A number of enlightening viewpoints habitually proscribe the consumption of definite foodstuff as well as show the way to foodstuff taboos. Such taboos are mainly frequent in primeval societies as well as they might, regrettably, confine the expenditure of animals foodstuffs so as to are fine resources of entire protein.

- In a region of mid-Africa, inhabitants consider so as to living things exploit is a disgusting organization discharge alike to urine. As a result, it is not extreme, in spite of its dietetic assessment.
- Several inhabitants in inaccessible areas of South East Asia keep away as of consumption of chicken or eggs for the reason that they are supposed to demolish human being productiveness.

Attitudes

An attitude is a method by which an individual sight amazing as well as an act towards it, regularly subsequent to appraising its advantage.

- ***Culture:*** For instance, consumption of raw animal protein as well as fish is a division of the Japanese society.
- ***Personal History*:** For case in point, being prepared to consume spinach the same as an infant might consequence in an important person keeping away as of all foodstuffs having spinach the same as a fully developed.
- ***Travel Experiences*:** For illustration, consumption of the cooking in foreign regions is a component of the educational understanding, excluding so as to cuisine would not automatically be consumed at dwelling.
- ***Perceived Status*:** (the location of somewhat concerning further stuff). For instance, filet mignon, lobster as well as truffles have an advanced rank than turnips, chuck steak, and flake fish.
- ***Economic Depression and War*:** For instance, older Australians as well as a lot of novel emigrants, who have gone all the way through chief gloominess as well as conflicts

by means of rationing as well as foodstuff scarcity, consume additional offal (living thing organs for instance beef liver as well as tongue), along with sense it is extravagant not to consume the whole thing provide on your serving dish.

Habits

A lot of the foodstuff options we formulate are custom. A custom is somewhat so as to we accomplish frequently devoid of thoughts. Foodstuff behaviors are similar as well as, similar to everyone behavior, are hard to break. Foodstuff behaviors, similar to the ones explained, are occasionally harmful furthermore require to be recognized if you wish for otherwise necessitate changing yours on the whole food. A few instances are:

- Take a wide coating of margarine or butter on your toast?
- Place tomato sauce on top of an animal protein quiche?
- Consume the identical breakfast cereal every day break?
- Imbibe soft drinks in its place of water?
- Shake over salty on foodstuff earlier than you flavor it?
- Purchase the similar category of bucks?
- Have covering on top of your emerald salad?
- Consume somewhat as almost immediately the same as you obtain habitat as of work or school?
- Consume somewhat charming by the side of the conclusion of every mealtime?

Figure 5.4: Psycological Habits Affecting Sensory Properties of Food.

http://factorsaffectingfoodselection.weebly.com/uploads/1/8/3/7/18371025/9237029.jpeg

Self-concept

Self-concept is an expression utilized to explain how we believe concerning ourselves (self-esteem) in addition to the technique we observe our individual manifestation counting the shape, size as well as the weight of our organization. Every of us has an image in our brain of:

- How we glance actually — weaknesses as well as strengths
- How fine matching we are
- Whether otherwise not we are elegant
- How shy otherwise sociable we are
- How fine we direct our wealth
- Whether otherwise not we have fine outfit common sense
- How brilliant we are at dance, music, art
- Our significance the same as a brother otherwise sister, son otherwise daughter.

Self-concept expands for the duration of childhood as well as it is appealing so as to this days' displeasure by means of body illustration enlarges as of the period of 11 years to the middle twenties, particularly in females. As well, our self-concept is not stationary; it might modify steadily over instance otherwise even quite a lot of occasions a daytime!

PHYSIOLOGICAL FACTORS

Physiological aspects influence the bodies required along with yearning in favor of foodstuff. If the organization is to stay fit as well as gathering properly it has to have sufficient concentrations of foodstuff having the necessary nutrients.

Hunger

Hunger might be definite the same as so as to emotion of weakness, emptiness, or pain founded by means of being short of food. It happens to additional strong the same as instance passes, in anticipation of we are capable to believe of slight as well however foodstuff. Designed for persons suffering severe lack of food, mitigating the organization of this hurt is the center of existence. Hunger is prescribed through a minute gland at the base of the head known as the hypothalamus. The hypothalamus has an amount of occupations in the organization, as well as machinery directly in the midst of the pituitary gland to:

- Manage body temperature
- Control thirst, appetite as well as body organization fluids
- Provoke wakefulness as well as sleep
- Organize the discharge of enlargement as well as sex hormones as of various glands all through the organization

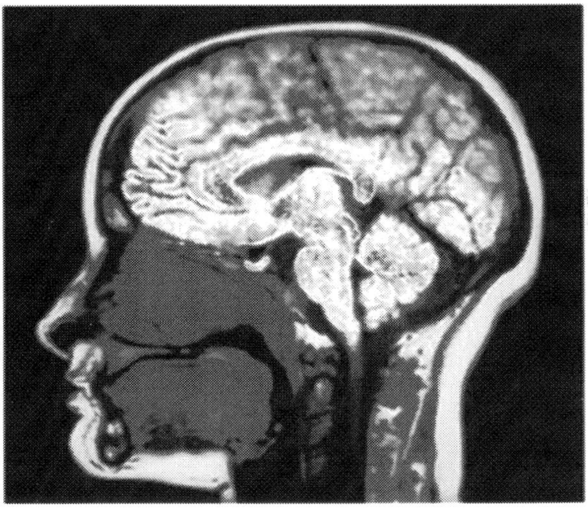

The hypothalamus (highlighted in green) is a small, specialised part of the brain that, among other things, lets us known when we are hungry.

Figure 5.5: Physiological Factor, Hunger Affecting Sensory Properties of Food.

https://encrypted-tbn0.gstatic.com/images?q=tbn:ANd9GcRLflSpW4gVVfUS1WhyBvkJFntpn0qgwWCLjjy4tCjHJnJDIruy

Appetite

Appetite is the yearning in favor of foodstuff still while the organization is not starving. Appetite might be activated by means of the display of mouth-watering foodstuff, the odor of foodstuff under processing, as well as still the reference to the foodstuff in the discussion. The hypothalamus records these cues as of the mind as well as launch messages to the head which persuade you to feel concerning foodstuff. The salivary glands are enthused as well as generate additional fluid: your oral cavity then commences to water. Dissimilar lack of food, if hunger is not contented it will finally absent

Nutritional Necessities

A lot of us choose foodstuff, which is healthful as we recognize so as to we will experience as well as stay fit. The foodstuff we consume ought to supply necessary nutrients so as to the organization is able to take up, as well as metabolize. There are five different groupings in favor of dietetic necessities they are organization magnitude or else level of activity, gender, category, age as well as physical condition rank.

Figure 5.6: Nutritional Necessities Affecting Sensory Properties of Food.

http://factorsaffectingfoodselection.weebly.com/uploads/1/8/3/7/18371025/5897246.png

Organization magnitude or category is the dietetic supplies of dissimilar sized individuals fluctuate. Persons who have superior construct necessitate additional nutrients to preserve as well as maneuver their organization developments. Correspondingly, those with a less significant organization magnitude necessitates a smaller amount of protein in favor of the preservation as well as renovate of organization tissues as their organization weight is fewer.

Age

Individual organization experience particular increase period all the way through verve. An infant's organization has a blown up skull, as well as the arms along with legs, are little concerning the respite of the organization. Early days sees spectacular modifications in organization magnitude. The arms enhance in length as well as muscle quality, which permits in favor of superior association as well as organization, as well as the legs, expand to formulate up and about half the body's altitude. The interlude of enlargement, as well as enlargement, prolongs all the way through teenage years in anticipation of the organization accomplishes a supplementary fully developed variety. For the reason that nutrients perform unambiguous occupations surrounded by the organization, a number of nutrients required by means of an entity are synchronized by means of the enlargement progressions.

Altitude of movement is indomitable as a result of a human being who is actually dynamic requirements to devour supplementary power giving foodstuff than a human being who shows the way a desk-bound, less active existence. Sedentary human beings require less of every nutrient than a full of life human being. If human beings devour huge amounts of power, however, does not go approximately a lot, the organization provisions the surplus as adipose tissue, stout.

Gender

The sexual category of an individual as well terminates their nutrient necessities. Organic behaviors for instance menstruation as well as childbirth represent so as to women require having an advanced nutritional ingestion of calcium as well as iron. Men have an advanced section of muscle tissue on their organizations, so they necessitate a superior ingestion of protein than women. Sexual characteristics will also have an effect on the elements of the organization wherever adipose tissue is stored. Women are mainly to stumble on additional adipose tissue on their bust, upper arms, waist, thighs as well as hips.

Health Status

A person's condition of physical condition might show the way to devour definite category of foodstuff and/or a smaller amount of further categories of foodstuff. If for instance, an individual has high cholesterol, which might ultimately show the way to heart illness, they might decide on to decrease the quantity of living thing stout in their diet. A few illnesses so as to might be linked to reduced diet comprise pellagra, beriberi, scurvy, rickets, anemia and osteoporosis. Anemia is an insufficiency illness where the human being exaggerated is lacking in hemoglobin and/or red blood cells.

Reactions to Food

Our individual point of views has a chief manipulation on our foodstuff choice. Persons choose otherwise refuse foodstuff derived from their response to the food's substantial manifestation, its appearance, odor, as well as touch. We make use of our right mind to estimate the eminence as well as the application of foodstuff, as well as judgments based on this response is known as sensory perceptions. Good-looking colors, as well as the imaginative understanding of foodstuff, arouse the hunger as well as the salivary glands, making the need in favor of definite items for consumption so as to might not otherwise have been selected. Foodstuff preparers as well take care of the position of foodstuff on the serving dish for the reason that they are conscious so as to the artistic manifestation of the serving of food know how to develop otherwise slow down the hunger of those to whom it is provided. Various individual's foodstuff preferences are to some degree for the reason that they have a bodily response to particular items for consumption or else constituents. The responses show a discrepancy flanked by persons nevertheless might take account of vomiting, abdominal swelling, itches, diarrhea as well as wheezing, skin rashes, disturbed sleep in addition to headaches.

SOCIAL FACTORS

The societies otherwise cultures so as to inhabitants reside, in conjunction with the category of contact that organisms enclose with one another, communal features, manipulate provisions selections. The significance of traditions, taboos as well as beliefs is of very much importance in the selection of food and the media, because of measurement of daily existence, sway to formulate definite foodstuff options. The education and employment, the category of the standard of living, the family size as well as the significance of generosity in the public assembly are also significant while making foodstuff selections.

Traditions and Culture

Traditions are ways of life so as to be repeated at definite durations by means of an associate of a cluster otherwise the public. A lot of customs connecting to unique circumstances involve foodstuff. Social moreover festive events at all times entail foodstuff to several extents, as well as the food is frequently the center of the occasion. Family customs habitually turn around foodstuff, the same as executing main cultural as well as social traditions in nearly all societies, for instance:

- ***Chinese New Year*** — festivity with demonstration of particular foodstuff, for example, rice cakes
- ***Easter*** — an occasion of receiving and giving confectionery otherwise, decorated hard boiled eggs as well as chocolate eggs

Figure 5.7: Traditions and Culture Affecting Sensory Properties of Food.

http://factorsaffectingfoodselection.weebly.com/uploads/1/8/3/7/18371025/7449374.jpeg

Standard of Living

Generally, a way of life aspects with the intention of manipulates foodstuff range concern:

- ***Employment*:** What you select to devour might rely on the substantial stress of your occupation. For instance, production occupation necessitates additional power than operational in a cartridge stock up does. Energetic occupations have need of the employee to consume additional carbohydrate loaded foodstuff meant for power although inhabitants who do sitting job, a duty so as to need minute physical progress, necessitate being cautious not to overdose.

- ***Education***: Apparently, wiser selections appear as of comprising additional data concerning the opportunities. Inhabitants as well turn out to be enhanced up to date regarding foodstuff as well as nourishment selection by management curriculum, examination of a variety of television shows, understanding journals food cupboard hand outs as well as fast food advertising material. The superior knowledgeable an individual is the subject of the sustenance concentration of foodstuff, dietetic necessities as well as provisions grounding, the better the possibility of wiser foodstuff choices.
- ***Geographic Position***: The staple foodstuff of a state will mainly count on whether it can be developed particularly the climactic surroundings along with natural features. If the weather is humid, then rising pineapples, coconuts, sugar cane also bananas is ideal as well as this foodstuff will be more plentiful, cheaper as well as repeatedly used in a diversity of food preparation traditions. Weather influence not simply the kinds of foodstuff developed in a region however as well the foodstuff selections inhabitants formulate. Wintry weather is the period of thick soups and porridge, lunchtime pasta along with meat pies, hot desserts as well as warm drinks. Summer carries the yearning in favor of fresh, bright, light foodstuff — juices, smoothies, fresh fruit salads, crispy salads, cold quiches, barbecues and ice creams.
- ***Interests and Journey***: A large number of states are nowadays unfastened to tourists; the internet let us compose pay for as of distant areas; as well as deal contracts among nations have intended so as to chief actions in one component of the globe may be felt all the way through the respite of the globe. While wandering, we perform an extensive variety of foodstuff, a number of which we are fond off as well as look for upon recurring residence. Personal interests and the interests of close personal friends can also influence food choices.

- ***Family Organizations and Roles*:** The household makes up determines the selection, superiority along with the amount of foodstuff devoted in a feast. As, adolescent offspring who have extremely susceptible taste buds have a preference of not as much of spicy foodstuff, whereas mature inhabitants might have a condensed taste sense along with frequently like extra profoundly flavored, particularly sweet and salty, foodstuff. Individuals are fond of as well as detest are habitually the mainly significant aspect in foodstuff collection surrounded by a family unit. Food preparation to special nutritional requirements in the relations might indicate so as to additional concern requirements to be in use to get ready foodstuff in a gorgeous as well as pleasant method.

Community Relations

Foodstuff has long been a representation of generosity as well as friendship. When associates go into your residence one of the earliest things you accomplish is present them a little to imbibe and eat. Provisions assist to make a comfortable impression in which still a timid individual might be a component of the assembly as a result of busying themselves with getting ready otherwise plateful provisions.

Media

The media take part an immense part in the foodstuff collections we compose, in a prosperous country such as Australia. Publicity of provisions is all over the place, daily we are uncovered to thousands of encouragement images, manuscripts, as well as echo as of the radio, television, magazines, cinemas, bill boards as well as the internet. A large amount of the foodstuff publicized by the medium/media is lesser in dietetic significance than its unrefined otherwise a reduced amount of processed substitutes. Though, these foodstuffs are accessible the same as if they are the exceptionally sources of a happy and healthy lifestyle.

Examination Assembly

Approximately an organism's peers are inhabitants in the similar age group with the similar communal rank. For the duration of teenage years, the control of the examiner assembly is strongest. The requirement in favor of receiving makes adolescents devour when and what their associates consume more willingly than what their parents believe they ought to consume as well as what is effective nutritionally. Trying novel stuff is safer in an age group, as well as the distribution of foodstuff is an excellent method to find to identify community as well as strengthen friendships.

Dwelling Generosity

Relation amusing in the habitat surroundings is going additionally relaxed furthermore a smaller amount of common. Organism's lives are busier in addition to, by way of the growing variety of carrying away foodstuff eating places and channels, it is mostly easier to include others carry out the cookery. Parents who are working do not have the instant to shop as well as prepare in favor of a ceremonial dinner festivity; it is easier to exploit carry away fast or else set off to an eating place where the washing up is done as a result of others.

ECONOMIC ASPECTS

Cost of Food

Cost is one of the most important factors when we select food. It is particularly important for low-income families, students and the elderly; it is not uncommon for these groups to spend 33 per cent of their total income on food.

Figure 5.8: Cost of Food Affecting Sensory Properties of Food.

http://factorsaffectingfoodselection.weebly.com/uploads/1/8/3/7/18371025/4767862.jpeg?260

The cost of food varies according to:

- ***How much is bought at once*** — if there is storage space, large quantities are a better buy • whether the food is in season or locally produced
- ***How much processing the food has undergone*** — pre-prepared marinated chicken skewers or bags of salad cost more. Surprisingly, less processed foods, for example, freshly squeezed orange juice from the greengrocer, can also cost more than the regular product.
- ***Place of purchase*** — if chosen carefully, food bought at a market or roadside stall can save money, but there is usually no money back guarantee
- Store specials and purchase of generic brands (large supermarkets have their own label).

Available Resources

Time is a resource that commonly limits what we buy and prepare for meals, while our knowledge of and skills in cooking may become rusty over time without use. Resources can be interchangeable. If you have the money you can choose to go to a restaurant for a meal, but if money is short you can use your skills, knowledge and food ingredients to make a meal. Time and money can be used wisely by freezing and safely storing foods that are in season, on sale, or in larger quantities than can be used immediately. The microwave can defrost foods quickly and is cheaper to run than a conventional oven.

The Marketplace

The marketplace refers to the place where consumers purchase food. It may be the corner store, the small local supermarket, the suburban shopping center, or the buying and selling of goods on world markets. Generally, the smaller the selling venue is, the more expensive the food items are. Small stores buy their food items from larger stores or wholesalers and increase the selling price to cover their costs. Selling venue is, the more expensive the food items are. Small stores buy their food items from larger stores or wholesalers and increase the selling price to cover their costs.

Figure 5.9: The Marketplace Affecting the Sensory Properties of Food.

https://horizon-magazine.eu/sites/default/files/field/image/Market.jpg

Occupation and Finances

As discussed previously, the type of job a person does influence their food selection. The physical demands of the job and its social expectations are reflected in food choices. The income received from employment also determines the quality and quantity of food chosen. For many Australian families, and many other individuals throughout the world, economic factors are the most important consideration when purchasing food.

Figure 5.10: Occupation and Finances Affecting Sensory Properties of Food.

https://encrypted-tbn0.gstatic.com/images?q=tbn:ANd9GcSRqXu62-tj6e7w-tjw-F2uKjwgvVuhFdvwdeDPPCwUqFTtnmtQmEA

REQUIREMENT OF SENSORY EVALUATION

- Classy sensory cubicles the same as for each ASTM principles by means of prescribed warmth (20C – 22`C) as well as RH by the side of 40±5%.
- In cubicles, appropriate lights: fair otherwise covered.
- In favor of dissimilar foodstuff appropriate glass wares and equipments.
- On behalf of chemical, a physical investigation of unprocessed as well as a processed foodstuff in the laboratory.
- For estimation of appropriate region designed for processing of foodstuff illustrations.
- Models' appropriate cryptograms of usually, three numeral systems are completed.
- Model appearance by suitable technique.
- For assessment appropriate sensory panel associates.
- In favor of a numerical assessment of sensory data by particular software.

SENSORY METHODS BRIEF DESCRIPTION

Following are the sensory methods:

- Discriminative Descriptive
- Hedonic Preference

Discriminative Descriptive

These preferences are product oriented and are objective as well as investigative. These are helpful to understand the measure and quality of a characteristic. These are important to be aware of the dissimilarity and similarity in the illustrations. These are used for equivalence intentions. These are done by a small number of chosen qualified panel affiliates.

Hedonic Preference

These preferences are inhabitants oriented. These are subjective and effective. These preferences play an important role in the approval of an item for consumption. In these preferences the primary impersonations are significant. These show the individual response or these likes substances. By and large additional panel associates, for example, illustrations experienced by means of inhabitants visiting a cubicle within a packed shopping center.

Explanation of Sensory Methods

- Discriminatory tests
- Simple differentiation - "which sample is different?"
- Triangle test
- Duo-trio test
- Directional dissimilarity - "which sample is sweeter?"
- Paired comparison test
- Preference tests: ranking test
- Scaling or scoring- "how 'pungent' is the sample?"
- Sensitivity tests – If a mossy odor is apparent?

Discriminatory Tests

Discriminatory tests are made on the way to

- Determination of whether the two illustrations are observable dissimilar.
- While 2 examples are chemically dissimilar along with might be supposed the same as matching.
- When there is a modification in the handing out technique of a food item for consumption.

These tests are enormously helpful in favor of estimation of a 'novel, enhanced 'edition of a previously accessible item for consumption.

Hedonic Test

For the measurement of foodstuff suitability, it is extremely useful. By the use of 9- point hedonic scale varieties as of 'extremely dislike' toward 'extremely like' the measurements are carried out. It is a community oriented examination through the big quantity of panel associates, mostly 50 as well as more than 50.

An expansion of hedonic scale through nine "infant pleasant" terminologies varying as of "high-quality" to "bad-quality" is employed among 5-10 year elderly offspring. A level containing "smiley" features furthermore achieves excellent reply in favor of foodstuff suitability as of offspring.

Hedonic Scale

The hedonic scale is the 9-Point scale that is explained below:

- Extremely like
- Very much like
- Moderately like
- Slightly like
- Neither dislike nor like
- Slightly dislike
- Moderately dislike
- Very much dislike
- Extremely dislike

Triangle Test

The panelist is offered through individual unlike as well as two similar illustrations. If achievable, the entire three examples ought to be accessible to the panelist on one occasion, as well as the panelist supposed to be inculcated to feel the examples as of left toward the right. The six potential organize mixtures ought to be randomized crosswise panelists. In favor of illustrations A as well as B, the six potential arrange groupings are BAB, BBA, BAA, ABB, AAB as well ABA. The panelist is inculcated to recognize the odd illustration along with confirmation of this respond

Duo – Trio Test

A reference is offered through 2 coded models. It is practical in favor of yields so as to enclose comparatively powerful odor, flavor, and/or kinesthetic sound effects. Whether otherwise not a sensory dissimilarity is really involved two examples determined by a Duo-Trio test. The extent or strength of distinctions is not extracted.

This technique is chiefly helpful:

- To establish whether an item for consumption dissimilarity consequence as of an alternation in dealing out, features, wrapping, or else storage space.
- To establish whether a general dissimilarity subsists, wherever no precise characteristics might be recognized like has been exaggerated.

Paired Comparison Test

Whether two yields varied in a particular characteristic, for instance, crispness, redness, and sweetness and so on are determined by the paired-comparison test. Concurrently two in different way coded illustrations are offered to every panelist. (30 and other associates) The panelist selects the one so as to be supposed like additional strength in the particular sensory characteristics.

The balanced evaluations associate the "enforced" selection as well as a result the panel of judges ought to offer a reply in every case. The prospect possibility linked through this investigation is 1/2. In preference examination, it is commonly applied.

Consequences Explanation

It is feasible to construct conclusion in relation to the yield under examination by application of numerical methods to the consequences. Most of the large FMCG corporations contain sections devoted to the investigation of the consequences and sensory examination.

A few software series are accessible these days in favor of the investigation of information created as of sensory examination. On the other hand, the numerical values will be significant just as the sensory estimation is performed by a suitable technique.

SENSORY DEVELOPMENTS

While the entire food items for consumption contain a variety of characteristics as well as magnitude, the factors typically considered in sensory estimation are

- ***Visual***: Wrapping manifestation, shape, color, item for consumption manifestation, dimension etc. Manifestation plays a significant role in the mixture to find out primary response to a foodstuff.
- ***Touch or Tactile***: Item for consumption texture, warmth, softness, feels and so on. The texture is evaluated in the course of physical contact as well as contact with foodstuff. The confrontations to masticate as well have an effect on touch, e.g. chewiness, crispy.
- ***Gustatory***: Flavor of the item for consumption. After food is to be found in the entrance, the exterior of the tongue, as well as supplementary susceptible covering responds to the feel of the exterior of the item for consumption in addition to generates a mouth-feel of the foodstuff. The oral cavity as well senses the warmth of the foodstuff that plays a significant motivation, for instance, hot soup, warm toast, cold ice cream and so on.
- ***Olfactory***: Taste as well as the odor of item for consumption.
- ***Auditory***: While consumption, echo– crunchy, crispy and so on.

To carry out a sensory investigation of foodstuff, there are a variety of processes.

- Founded on the investigation process, models are formed as well as board members are selected.
- The sensory development is complex as well as entails the observation of the entire senses as foodstuff is consumed.

NEEDS OF SENSORY EVALUATION

Sensory evaluation is associated with some needs and these needs are following:

- Within a collection of foodstuffs observation of resemblances or dissimilarities.
- Estimation of on hand food item for consumption in opposition to a standard example.
- Investigation of foodstuff model in favor of supplementary developments founded on marketplace response.

- Extraction of definite reaction to a foodstuff example: whether satisfactory otherwise not with clients.
- Revision of a scrupulous belonging in a constituent or else a food item for consumption.
- Estimation of the food product, whether a prepared food item for consumption congregates its customary example or original requirement.
- Attaining the response information with the intention of making choices as well as perform appropriate adjustment within a food item for consumption.

USES OF SENSORY ASSESSMENT

Sensory evaluation is used for many purposes. It finds its uses in a wide range of sensory evaluation. Following are the uses of sensory evaluation:

- Development of new product
- Prior to a new product encouragement
- Establish the effect of formulation variations, particularly when accessibility of natural ingredients is insufficient.
- Understanding the effect of processing variations.
- Guarantee group reliability.
- Examine shelf-life alternation.
- Estimate consumer approval.
- Specialist versus consumer sensory information.
- Sensory feature power as well as consumer reliability.
- Sensory assessment in Food disciplines' courses.

SENSORY EVALUATION OF SPICES

Because of the unique color, sharpness as well as the flavor of the spices, these are crucial foodstuff constituents. Volatile oils instruct the exclusive odor, as well as oleoresins, offer the flavor of spices. Spices are used during preparation of medicines as well as possess several pharmacological and medicinal properties. The odor mixture in flavor so as to create the flavorings exploited in the provisions production are categorized into functional groups like thiols, aldehydes, ketones, terpenes, ethers, amines, esters and additional miscellaneous mixtures.

SENSORY OBSERVATION OF MIXED MASALAS

Sensory estimation of mixed masalas used is a combined task of taste, appearance, texture, odor, and flavor. Complex mixtures of spices are mixed masalas as well as spice blends accessible for exploit. It is essential to have a perceptive of the unit report of the flavors used in a blended masala. Sensory examination of blended masalas necessitates a definite intensity of proficiency to notice the portions as well as proportions of a mixture of spices utilized in the synthesis of the mixture.

Generally, in blended masalas, the observations of the occurrence of a unit flavor rely on:

- The grade otherwise sort of a spice found, e.g. high Pinene concentration in black pepper generates a turpentine akin to fragrance while Caryophyllene rich black pepper oils offer charming fruity note.
- The major volatile constituents as well as the ratios in which they found in the mixture.
- In a masala mix, the shape of ripeness of the spice at the time of exploit; e.g. a typical odor is practical if undeveloped coriander seeds, as well as its vegetative components, are exploited in a mixture because of an aliphatic aldehyde in the coriander oil while a pleasurable, as well as charming aroma, comes as of the mature, grown-up seeds of coriander by reason of the occurrence of a monoterpene alcohol known as Linalool.
- In the masala, the portion of the flavor extract used; e.g. the distinguishing flavor of ginger is supposed when the condensation unstable oil constituents such as monoterpenes, are found while the oral cavity sense and pungency is felt because of the occurrence of the no steam volatile concentrations of ginger, known as gingerols.
- The degree of the entire spice against the extracts as well as other volatile parts utilized for improvement at the period of production of the blended masala will influence the sensory observation of the flavor mixture.

SUMMARY

According to the inner circumstances, an occurrence as well as specific time, individuals choose as of their accessible foodstuff supplies the substances that are mainly satisfactory, gratifying as well as practical. By means of the refusal of unidentified, injurious or else out of favor foodstuff, this is heralded. The aspects manipulating provisions assortment might be randomly categorized like sensory, cognitive, in addition to physiological.

At any rate, the physiological aspects include two chief utilities, in the outline of accessible information. One is to make sure bring in meant for the body system of the two vital nutrients, carbohydrate along with protein; further is to shield the body organization as of dangerous materials. According to extent of detest and fondness, the sensory aspects classify foodstuff. The cognitive features monitor the sensory indications as of the viewpoint of full of health and make a possible variation of food-related performance to the common surroundings.

CHAPTER

6

CLEANING AND DISINFECTION IN FOOD PROCESSING OPERATIONS

INTRODUCTION

Foodstuff protection organization mostly put HACCP by the side of the center of manage in favor of precise foodstuff dangers. This formulates good judgment chiefly everywhere a definite vulnerability associated in the direction of unfavorable contact in favor of the customer might be recognized. On the other hand, no substance how forceful the HACCP organization might be it requires a reverberation curriculum of vital powers so as to take in hand broad-spectrum vulnerabilities, several of which might not be acknowledged. These courses are illustrated the same as Pre-requisite Programs (PRP), Control Points (CP), as well as Good Manufacturing Practices (GMP) to recognize a few. In spite of the phrase described they entirely symbolize the essential necessities in favor of excellent performances to offer protected surroundings in favor of the assemblage of foodstuff. Amongst the central of these is the requirement to hygienic as well as disinfect your deposit as well as tackles adequate to raise foodstuff free of substantial, microbiological, allergenic and chemical risks. As well, it is significant so as to workers recognize the cause why a foodstuff fix has to be hygienic. Basically, inculcate inhabitants is infrequently sufficient to sustain high principles, they must as well recognize the explanation why, counting:

- To decrease the dangers as of foodstuff dangers, foodstuff extermination as well as distant body organization contagion.
- To observe by means of local as well as worldwide direction.
- To assemble precise consumer necessities for example, Tesco.
- To congregate the supplies of worldwide foodstuff protection principles.
- To sustain optimistic inspection as well as examination effects.
- To allocate utmost deposit efficiency.
- At hand a sterile illustration.
- To encourage protected operational situations meant for employees, out workers as well as companies.
- To preserve item for consumption shelf-life.
- To evade insect influx.

By the side of the primary fundamental intensity, the illustration manifestation of a foodstuff industrial unit is a suggestion of the principles as well as traditions of the corporation. It has a powerful contact on the observation of an examiner otherwise guest as well as might manipulate on the whole product of inspections as well as protecting the original industry. In support of this explanation, the illustration sanitation of a corporation is the same as significant as comprehensive HACCP strategies.

It is frequently supposed the same as an essential malevolence that does not include worth to an artifact unswervingly. The charges of cleaning as well as indeed the expenditure of not crackdown are infrequently calculated normally as a result of foodstuff companies. The characteristic charge fundamentals of a clean-up series comprise:

- Employment as well as regulation.
- Water management, supply in addition to procuring.
- Monitoring.
- Warmth water.
- Downtime.

- Cleaning tackles.
- Substances.
- Corrosion.
- Affluent.

Of these, manual labor is usually the major feature counting for more than 60% of the entire cleaning financial plan whether resourced below agreement otherwise domestic. When charge forces appear too exposed typically work is usually cut down. Although this might accumulate capital in the temporary, in excess of the moment it will direct to an amount of roundabout expenditures counting a decrease in shelf life, raise in item for consumption recalls, complaints, and authoritarian limitation as well as a loss of corporation. The feasibility of the industry will finally be impacted. The subsequently most noteworthy expenses are water as well as chemicals which might differ conditional on the foundation as well as dealer.

CLEANING AND DISINFECTION: IMPROVING FOOD SAFETY AND OPERATIONAL EFFICIENCY IN FOOD PROCESSING

Cleaning in the foodstuff business is not an uncomplicated duty. On the other hand, it is a serious step inside foodstuff assembly as it is essential to sustain as well as security to foodstuff safety. Considerate a range of earth dispute, why we clean as well as how disinfectants, as well as detergents work, are the solution to ensure a protected, clean developed situation.

- **Prevent Products or Ingredients Transfer**— When numerous foodstuffs are made on the similar engine, it is unwanted to cross pollute substances or else exchange as of one item for consumption to the subsequent.
- **Avoid Microbial Contamination**— This might direct to numerous disturbances, less manufactured goods feature, damage to fitness or else still verve frightening conditions in a few cases. Cleaning only is no assurance of distillation, except it is a precondition to disinfection.

- **Guarantee Disinfectant Effectiveness**— Soil forces the efficiency of a decontaminator. The smaller amount soil on the exterior, the most useful sanitize will be by the side of decreasing microbiological contagion.
- **Improve Plant Effectiveness**— Soil infectivity decreases the effectiveness of apparatus as well as the manufacturing development.
- **Enhance Protection**— Services so as to be not uncontaminated efficiently include additional potential protection dangers, similar to falls as well as slips, because of foodstuff dissipate on the ground. As well, chief occurrence by reason of increase of top soil in apparatus might furthermore take place.
- **Impact Economic Suggestions**— Minimizing dissipate as of spoilage might considerably enlarge the verve of apparatus as well as equipment.
- **Reduce Lawful Consequences**— Even though it might not be widespread information, there are frequently officially permitted supplies in favor of foodstuff conveniences on the way to the hygienic exterior as well as apparatus on the way to a definite principle.
- **Increase Stakeholder Assurance**— In conclusion, the manifestation of deposit as well as property is frequently unnoticed on the other hand the psychosomatic confidence as well as benefits achieved as of sanitary, disinfected apparatus, as well as organized environments, have a momentous effect on mutually employee pleasure as well as consumer assurance.

CLEANING

Cleaning as well as disinfection ought to be measured as two distinct points within the cleaning process. Disinfection is the inclusive subtraction of remains as well as top soil as of exteriors, parting them visually hygienic as a result so as to consequent cleaning will be efficient. Devoid of efficient cleaning, disinfection will be cooperation.

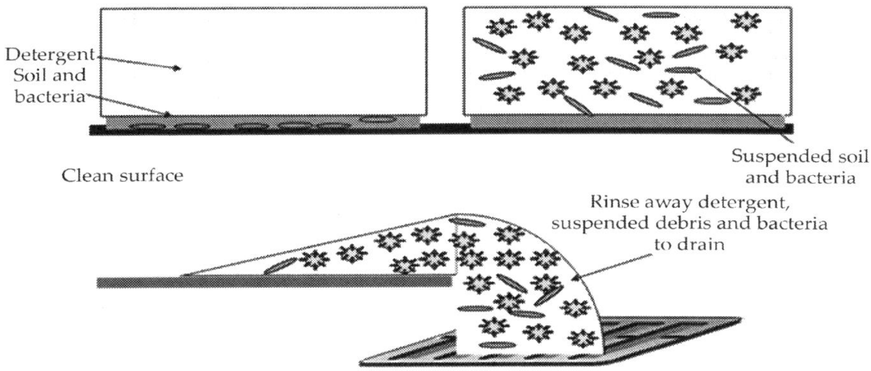

Figure 6.1: Cleaning.

http://www.foodsafetymagazine.com/fsm/assets/Image/CleaningFig1.jpg

Detergents are in the habit to get rid of top soil as of an exterior. The soil is a combination of foodstuff misuse as well as microorganisms is on otherwise close to the exterior of the handing out apparatus, walls or floors. The act of the detergent solution is to suspend this top soil as well as microorganism's combination far as of the exterior as well as let in favor of it on the way to be washed off en route for the exhaust. Though, there are several soils present in the foodstuff business as well as the cleaning process as well as detergent used so as to obtain the preferred detergent exploit is dissimilar in favor of every top soil.

The most widespread top soils, carbohydrates similar to starch, cellulose, and sugar are the simplest to get rid of. Proteins, milk, eggs, and meat are in all probability the most complicated as modifications in heat as well as pH modify the formation of the protein moreover connect it to extra molecules, raising their persistence as well as frequently representation them unsolvable. For case in point, as milk is soluble in water, when you in excess apply warmth to a pan of milk, the consequential milk top soil forms complicated to get rid of the saucepan.

Fatty oils are not water soluble as well as cause a superior confront than carbohydrates. At this point, it's essential to make use of alkaline cleaners as well as eminent warmth over the melting peak of the stout to attain a well-organized sanitary. Mineral salts, the inorganic foodstuff top soils, cause scale development on apparatus. Acidic cleaners are necessary to proficiently get rid of the scale.

There are four variables in the clean-up development so as to may affect its effectiveness to do away with top soil:

- Detergent and Concentration
- Time
- Temperature
- Physical Action

In 1959 develop by Dr. Herbert Sinner; Sinner's Circle is generally called the form to display so as to minimize one of the four aspects might be rewarded as a result of mounting an additional. In favor of case, you might be capable to enhance the warmth to make possible for you to make use of a minor amount of substance.

DISINFECTION

Disinfection is the procedure as a result of which microbes are inactivated therefore so as to their numbers are decreased to an intensity which is neither injurious to physical condition nor to the feature of unpreserved goods. Subsequent clean-up, exteriors will be liberated as of top soil other than microbes stay behind. By means of corroborated de-contaminators on top of exterior, follow the information as well as get in touch with the company of times, decreases microbes levels in the direction of the mandatory intensity in favor of foodstuff manufacturing

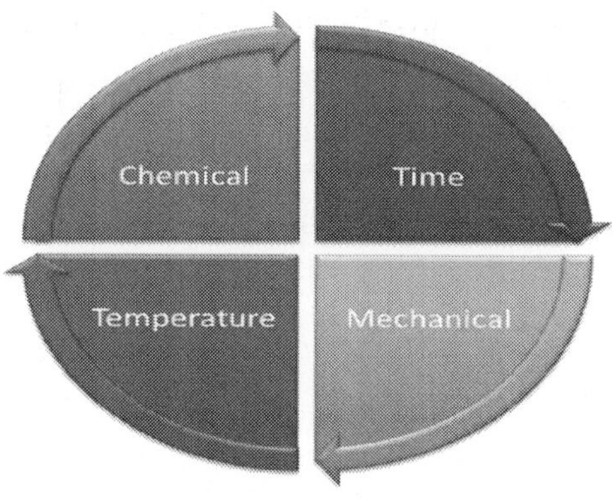

Figure 6.2: Cleaning and Disinfection: Improving Food Safety and Operational Efficiency in Food Processing.

http://www.foodsafetymagazine.com/fsm/assets/Image/CleaningFig2.jpg

The technique through which de-contaminators execute the microbes, referred to the same as their "mode of action", fluctuates in the midst of the dynamic constituent.

While choosing a decontaminator, numerals of contemplations necessitate being prepared. Applications, counting the compatibility by means of equipment originate within the area being sanitized; temperature requisite, the impact of water hardness, necessary concentration, toxicity, available remains.

CLEANING VALIDATION

The validation of cleaning as well as assuring principle, consistent results has become a topic of elevated priority into the food processing business. This validation is used in the direction to demonstrate proof so as to the clean-up system consistently will perform as expected. An exterior is chemically clean if there are no microscopic residues of top soil residual as well as no residual detergents or else sterilizer substances on the way to contaminate the food product. Determinations of chemical hygiene require tools except than the individual observation. Techniques in favor of determining such small amounts of top soil consist of:

- **Visual moisture on surfaces**— Soiled exteriors become hydrophobic, as a result water will bead up as well as show the presence of top soil.
- **Dyes**— Definite dyes by means of an affinity in favor of certain soils similar to starch or else protein might as well be applied to an exterior to formulate the top soil observable.
- **Optical methods**— These are used quite regularly inside various food processing procedures meant for detecting top soil. Interruption of a brightness beam reflectance, as well as absorbance is the entire optical means of detecting soil.
- **Adenosine triphosphate (ATP)**— ATP concentrations are present in the whole organic substances. Determine its presence is a very effective way of providing instantaneous response on the way to determine the hygiene of serious organize region.

The basic principles discussed here should provide insight into the variables that should be taken into account when devising a cleaning schedule. Controlling these variables will present opportunities to deliver real improvements in operational efficiency in food processing while improving overall food safety.

OUTLINE

Cleaning is the network to create, a place otherwise item, getting rid of any dirt otherwise everything harmful on the way to the hygiene.

Disinfection is the removal or else the inhibition of microbes within a place, a thing or else on the exterior plane of the individual organization. Within foodstuff manufacturing, foodstuff, clean-up as well as disinfection are considered in the direction to get rid of dirt as well as obliterate the microbes found in the piece of equipment as well as wrapping. In several cases, these processes ought to be accomplished on the foodstuffs themselves unhygienic as a result of used for soil, microorganisms, or else remains of insect repellent.

Cleaning along with disinfection procedures are in the midst of the mainly significant of the IAA as well as this in favor of a variety of causes:

- The value of the refined item for consumption is frequently inclined as a result of the flavors of strangers by reason of microbial expansion. These expansions are maintained by the remains of this item for consumption within the piece of equipment or else in a pot after utilization, or else as of deposit so as to be created all through the handing out of definite foodstuffs for example milk, beer and so on.
- The capability on the way to high temperature action is extremely maintained by the preliminary inhabitants. To be sure, as disinfects pasteurizes or else on an item for consumption as a result of warmth, instant of handling relies on the primary microbial consignment. It is greatly longer if this weight is not summarized by means of an introductory cleaning. This enhances the charges of action as well as might reduce the eminence of the item for consumption.
- The dust might have pathogenic microbes as well as, consequently, comprise a cause of infectivity is tremendously damaging on the way to the foodstuff.

- The existence of remains for example desiccated crusts otherwise distorted pests or else their larvae, rodents otherwise still inside the pack up foodstuffs might have a disastrous power on the evaluation of the customer.

Nature of Infectivity

It indicates dirt otherwise dirt products are completely different. These products are constituents of foodstuff additional otherwise less tarnished or else customized by means of humidity, heat, oxygen, light, cold and /or microbes. These mechanisms might be extra or else less merged by means of others: filter molds otherwise yeasts, plant minerals remain or else, kernels of the environment or else completed with organization.

In speculation, a blemish is differentiated by means of the proportion of its water solubility as well as lipid solubility; in addition to it is this feature so as to ought to be taken into explanation while selecting substance sanitization means. The blot is a multifarious natural substance, inorganic substance as well as microbial microorganisms. Apiece of these three families of complexes ought to be recognized in point.

Organic compounds are lipids, carbohydrates and proteins. The inert complexes might be separated into numerous groups that are derived from their power and nature of removal: The alkaline earth aluminates, salts, oxalates and silicates.

Microbial microorganisms are categorized into two classes: helpful microorganisms as well as damaging microorganisms. When found concurrently in the surroundings of the foodstuff item for consumption or else the item for consumption itself, it is impossible in present application, in the direction of to formulate discerning handling, on the way to manage microbial verve within an item for consumption, we initiate by means of devastating the whole thing as well as coatings practical devastated.

Cleaning Products

Detergents are manufactured goods whose solution helps en route to get rid of dirt or else supplementary unfamiliar remains polluted exteriors. In anticipation of the 1940s, soap was the single significant detergent. Nowadays, there are a lot of others.

Method of Action of Cleaning Products

As you place a detergent in water, it considers so as to it is soluble in spite of its amphiphilic nature. Actually, it is an assemblage of detergent molecules in the appearance of micelles so as to stay behind poised inside water. Inside micelle the parties' polar, hydrophilic top on the way to the outer, in make contact with water as well as the parties non-polar, hydrophobic is touching inwards, get in touch with additional hydrophobic fractions.

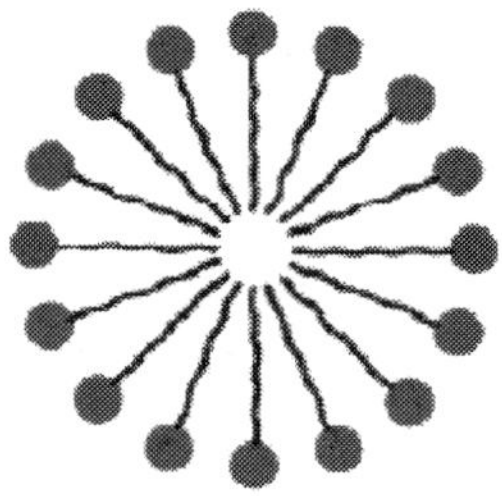

Figure 6.3: Detergent Micelle.

http://www.azaquar.com/sites/default/files/doc/en_images/chemistry/ca_lipides12.gif

As agitated water surrounding **a** surfactant, hydrophobic top soils, for example, grease, oil and so on, related to the inner component of the micelles because of their hydrophobic character. As a result, contaminated are micelles in suspension as the residues of the categorization of molecular magnitude, 100 to 200 angstroms. While soaking, these micelles are driven by means of water as well as dirt is divided as of their carry.

On the other hand, the development of micelles is experiential in anticipation of the significant micellar concentration, application on which the molecules shape micelles. The detergent actual act takes place so as to on meditations of surfactant over the significant micellar amount. In favor of alkyls-aryl-sulfonate, the important micellar level is almost 0.1%, although it is powerfully influenced as a result of temperature as well as the existence of saline, furthermore static, might considerably reduce the significant micellar amount as well as, as a result enhance the detergent.

Cleaning Products

Surfactants

A surfactant is by means of categorization an organization capable in the direction of modifies considerably the characteristics of the exterior or else boundary solutions. On the whole, these organizations are known as amphiphiles, i.e. have one or else additional polar groups as well as non polar hydrocarbon series usually having 10 to 20 carbon atoms.

One category in general surfactants in relation to the polar character of the assembly:

- The anionic surfactants containing a polar acid assembly, the chief yields of this sort are:
  - Soaps or else alkaline salts of fatty acid,
  - Alcoylsulfates the primary as well as secondary
  - The alcoylaryl sulfonates.

These yields are outstanding detergents; they are not bactericidal as well as, apart from in favor of the serialization, are not extremely susceptible en route for water hardness.

- ***Cationic surfactants*:** These are solubilized on the way to a quaternary ammonium collection or else an ethoxylated amine. The previous might be outstanding bactericidal as well as the concluding are of deterioration inhibitors into acidic medium.
- ***Surfactants, anionic non-condensate*:** These are normally of ethylene oxide on alcoylphénoles or else fatty alcohols. These yields are outstanding bactericidal detergent whose solubility is superior in cold than hot, as well as this solubility amplifies by way of the integer of molecules of ethylene oxide. An outcome of the series solubilizing cannot compose fluff.
- ***The amphoteric surfactants*:** These are solubilized by means of together cationic as well as anionic assembly as well as a little of these yields have a limited uses as bactericides.

Further Cleaning Products

Strong alkalis are outstanding bactericides as well as detergents. For instance, caustic soda, which as well has sporicidal as well as virucidal action is used when the resources comprising the machinery allows. Strong acids are excellent bactericides as well as detergents. We make use of hydrochloric acid, phosphoric as well as nitric. Sulfuric acid compliance has outstanding bactericidal, still alongside the spores.

Weak acids for instance gluconic acid, citric, sulfamic perform a cleaning exploit significantly subordinate than so as to of strong acids as well as their bactericidal exploit is basically nothing, it is as well alkaline salts whose exploits might in no method be measured up to so as to of caustic soda. Chlorine in general observed like a de-contaminator in addition has characters exploited in favor of the exclusion as a result of oxidation of proteins.

Mode of Action of Chemical Agents

The bactericidal exploit of chemical agents might result as of a modification in permeability of the cell membrane, or else a splitting of this membrane, exploit taking place on definite enzymes or else definite cellular proteins, as a result of oxidation, perform a decrease or else hydrolysis of essential ingredients, or else of intervention through necessary substrates.

The Products of Disinfection

The selection of yields in favor of disinfecting as well as cleaning equipments relies chiefly comprise the equipment, the experience of decomposition still in performance an important character.

Several clean-up yields previously discussed above have significant bactericidal exploit. They comprise quaternary ammonium yields, as well as strong acids along with alkalis; they might be active still in opposition to microbial viruses as well as spores. The dilemma of option of supplementary bactericidal agents takes place as a result so as to in cases where the supplies do not sustain strong alkalis as well as strong acids, as well as we want to develop cleanliness as well as avoid re-infection after action.

Bactericidal agents are the for the most part generally exploited halogen in exacting chlorine as well as hypochlorite's along with these yield are dynamic by the side of amount of chlorine as of a small number of ppm, parts per million. It furthermore exploits the item for consumption of co-crystallization of CLON as well as having roughly 3% of active chlorine in the structure comparatively steady.

Chlorine as well has the drawback of being irreconcilable among the surfactants by means of which it acts in response through storage that is why it was anticipated long before, the exploitation of combinations of iodine as well as agents surface non-ionic, condensates of ethylene oxide. These yields chosen in the indication iodophors have a sporicidal as well as bactericidal activity still at levels of a small ppm of liberated iodine on pH less than 4.5. Iodine trichloride ICl_3 has the benefit of being well-suited through anionic surfactants in the direction of to be extra stable, not as much of corrosive as well as in the direction of to be bactericidal on pH neighbor's impartiality.

As well exploited the same as bactericidal agents, formaldehyde, the silver ions, ozone, hydrogen peroxide as well as peracids, SO_2, the 3-5-4 'tribromo-salicylanilide, as well as sulfates and so on. In support of sterilization of wrapping the exploit of concentrated solutions of warm hydrogen peroxide is common. It as well exploits phosphoric acid, chlorine as well as per acetic acid.

In favor of sterilizing foodstuff item for consumption themselves has anticipated the utilization of ethylene oxide combination by means of an inert gas. Sterilization warmth rise might be extremely efficient in the case of machinery specifically premeditated in favor of the sterilization by means of warm water or else by means of vapor condensation. In the concluding point, the condensed water as well as air ought to be get rid of by means of efficient traps located intelligently

PRACTICAL REALIZATION OF CLEANING AND DISINFECTION

Cleaning is the elimination of leftovers of foodstuff leaves, microbes, as well as distant organization, several pests, and so on. In standard situations, the elimination of microbes as well as spores is not inclusive as well as the cleaning ought to be pursued as a result of disinfection. A key difficulty is deciding whether on the way to carry out the cleaning as well as disinfection within one step or else in two different as well as consecutive processes.

As soon as the pH of the item for consumption is below 4.0 it is not essential on the way to demolish spores of Bacillus as well as Clostridium, as well as might, as a result, make use of the yields extremely effectual quaternary ammonium as well as non-corrosive. The same as in opposition to so as to the item for consumption has a pH higher than 4.5 ought to be most frequently employed halogen. Cleaning as well as disinfection of the pipes are completed as a result of driving the most excellent solutions on the way to attain adequate confusion, it ought to be illustrious so as to by the side of identical rate, the Reynolds numeral fluctuates the same as the converse of the distance of the pipe; any extension communicate on the way to a decrease of commotion essential on the way to achieve acceptable consequences. It is furthermore essential on the way to comprise in the route, the sterilization of openings by three-way control device by means of a sensible rotary motion as well as avoids whichever region wherever the Reynolds numeral is below than 6000.

Cleaning as well as disinfection of tanks as well as vessels cannot normally be attained as a result of progress as a result of unnecessary quantities of clean-up solution as well as because of inadequate commotion all through the conversion of these solutions.

The procedures of accomplishment are:

- **A brush:** This organization is profoundly reliant on the human aspect as well as ought to be unconcerned at any time achievable.
- **Jet manual:** The persuasion of the individual reason is as well prominent so as to this scheme is a development greater than the preceding organization.
- **Jet hot:** This is further efficient than low-temperature procedures, presented so as to the table is not only on elevating the warmth on the way to attain the cleaning as well as disinfection.
- Jet as of the ball or else cylinder punctured.
- Jet as of automatically proscribed turnstiles.
- **Jet foam:** An item for consumption in favor of cleaning as well as disinfection ought to be added a foaming agent so

as to allow extremely extend a coating of fluff lying on the apparatus. It is these bubbles permit in favor of the preferred instance as well as after that gets rid of it as a result of simple cleaning.

- Containers are generally cleaned as well as disinfected by a particular procedure. This is the casing in favor of glass bottles experience soakings injections as well as solutions in the direction of in the order of 1% NaOH to 70 ° C. This solution is in general additional probably a bactericidal driving force.
- Cleaning as well as disinfection of foodstuff yield is regularly completed by means of by way of rotating strategies wherever they are subjected in the direction of jets of detergent solutions as well as bactericides.
- In manufacturing the construction of metabolites, fermenters are frequently listed in favor of routine cleaning as well as exchange jets as well as sterilization below force of water vapor.
- Rinsing is the elimination of bactericides as well as detergents by means of water. The difficulty so as to be contrasting on the way to optimizing this procedure as a result of minimizing the amount of water devoted as taking away traces of detergent as well as disinfectant. The existing style in the globe, subsequent to the utilization of cleaning agents as well as sufficient cleanliness, not to clean on the way to evade re-infection.

STEPS OF CLEANING AND DISINFECTION

Cleaning as well as disinfection is able to edge contains the subsequent steps:

- Rinsing
- Cleaning with strong base
- Subsequent rinsing
- Acid washing
- Subsequent rinsing
- Disinfecting
- Final rinsing

Such a procedure is extended, expensive as well as consequences within the development of watery sewage within significant magnitudes. It is consequently significant on the way to accomplish or else so as to decrease the shortcomings of this technique as preserving the identical altitude of efficiency.

Inside the confines of the probable stages in proportion to the alkaline as well as acid cleaning might be mannered within one step by means of an item for consumption so as to do away with mutually the natural as well as inorganic stains. Therefore, the procedure of cleaning as well as disinfection by seven steps might be attained in five steps:

- Rinsing
- Chemical wash
- Subsequent rinsing
- Disinfection by employing a chemical compound
- Final rinsing

Correspondingly, the stages of cleaning as well as disinfection might in various cases be performed concurrently. Foaming equipment so as to guarantee to make contact with time longer between detergents/disinfectants as well as the surfaces on the way to be treated is mostly utilized. In this case, the procedure of cleaning as well as disinfection is mannered by three steps:

- Rinsing
- Cleaning as well as disinfection by means of a chemical agent
- Final rinsing

Cleaning as well as disinfection in three stages let in favor of a substantial increase in energy, time, water and labor. On the other hand, this procedure is merely efficient if the quantity of infectivity is on a comparatively low concentration.

FREQUENCY OF CLEANING AND DISINFECTION

Every sort of development and/or apparatus necessary is determined so as to the number of times the cleaning as well as disinfection will be essential, by an epoch of instance. The incidence of process relies on the character of the foodstuff item for consumption being applied as well as knowledge alteration.

The same as observes cleanliness, the superlative method to verify the occurrence of procedure is to perform a microbiological organize in the direction to verify the development of the microbial vegetation the same as a occupation of instance. This will subtract the instant by the side of which it ought to be disinfected.

CLEANING AND DISINFECTION

It is inadequate to decide a plan of cleaning as well as disinfection as well as on the way to be appropriate; however we as well ought to be guaranteeing its efficiency. This is reviewed as a result of subsequent pre-established controls so as to let one hand on the way to make sure so as to the curriculum of cleaning as well as disinfection is successfully applied as well as secondly to make sure so as to it is efficient.

Organization of cleaning as well as disinfection is attained by two customs: The primary is on the way to summarize occupation as well as illustration examination on the way to make sure the sanitation of machinery, premises, as well as apparatus, in addition to the subsequent is on the way to accomplish the microbiological examination in the direction to guarantee so as to the agenda of cleaning as well as disinfection is still efficient. Later than cleaning as well as disinfection, the microbial consignment of the exteriors is evaluated as a result of examining the exterior on the way to be examined by means of a disinfected wash which is after that transmitted within disinfected condensed water intended for dilution. The kernels are isolated by means of a Vortex mixer as well as the counting is completed later than beginning the culture intermediate as well as incubation by a specified warmth dependent on the category of microbes required.

Performance for microbiological examination as a result of cleaning or else as a result of gélosées blades has the benefit of being discerning as well as accurate if you yearn for. On the other hand, they are sluggish as well as cannot formulate a conclusion appropriately. The exploitation of the ATP metry meant for reviewing the efficiency of cleaning as well as disinfection the same as a substitute in the direction of these processes as well as has the improvement so as to it might be speedily recognize responsive regions in conditions of cleaning as well as disinfection.

- **Mechanical techniques** In favor of observing sterilization consist of reviewing the series time, pressure, as well as temperature of sterilization apparatus as a result of examining the gauges or else exhibits on the disinfectant. Various tabletop sterilizers have recording measures so as to print out these restrictions. Accurate understandings do not make sure sterilization, however inaccurate understandings could be the primary suggestion so as to a difficulty has happened by means of the sterilization sequence.
- **Chemical indicators**, interior as well as exterior, make use of susceptible substances on the way to review substantial situations for instance high temperature all through the sterilization procedure. Chemical displays for example high temperature susceptible strip alteration shade speedily while a specified parameter is accomplished.
    - **The Internal chemical indicator** ought to be positioned within each sterilization enclose on the way to make sure the sterilization mediator has go through the covering substance as well as in point of information arrived at the apparatus in the interior.

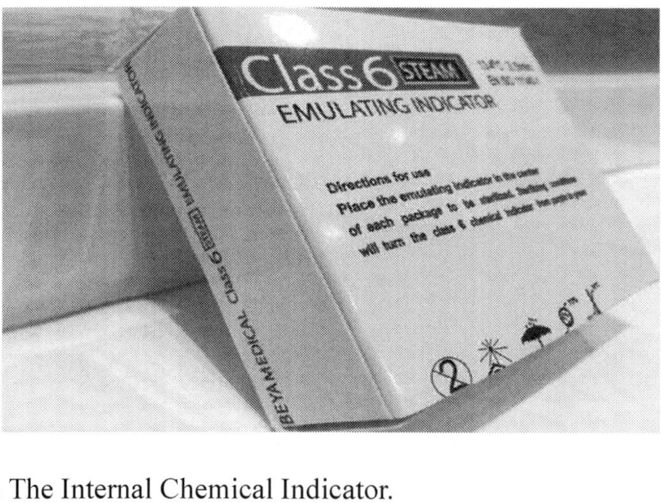

Figure 6.4: The Internal Chemical Indicator.

https://4.imimg.com/data4/DP/AR/MY-21474379/chemical-indicator-class-6-500x500.jpg

- **The External indicator** ought to be employed as the inside pointer cannot be seen as of exterior the wrap up. Single-parameter interior displays offer in order on merely one sterilization considerations as well as are accessible in favor of dry heat, unsaturated chemical, as well as vapor steam.

Figure 6.5: The External Indicator.

https://jody929lee.files.wordpress.com/2015/04/comply-1228-300.jpg

- **Multiparameter internal indicators** determine 2–3 parameters as well as might provide an extra consistent suggestion so as to sterilization circumstances have been assembled. Multiparameter inner pointers are simply accessible in favor of condensation sterilizers (i.e., autoclaves). Submit on the way to producer commands in favor of appropriate utilization as well as placement of chemical displays.

Indicator test consequences are exposed instantly subsequent to the sterilization sequence is entire as well as might supply an untimely suggestion of a dilemma as well as where the dilemma takes place in the development. If the internal or else outside display put forward insufficient handing out, the thing that has been developed ought to not be utilized. For the reason that chemical display do not demonstrate sterilization has been accomplished, an organic display, i.e., spore test is necessary.

- **Biological indicators** (BIs) are the mainly conventional sources of observing the sterilization development as they unswervingly conclude whether the mainly opposed to microbes, for example, Geobacillus otherwise Bacillus species are found more willingly than simply resolve whether the substantial as well as chemical circumstances essential in favor of sterilization are met. For the reason so as to spores made in BIs are more resistant as well as present in superior information than are the universal microbial contaminants institute on patient care apparatus, an inactivated BI specified so as to additional possible pathogens in the consignment have moreover been exterminated.

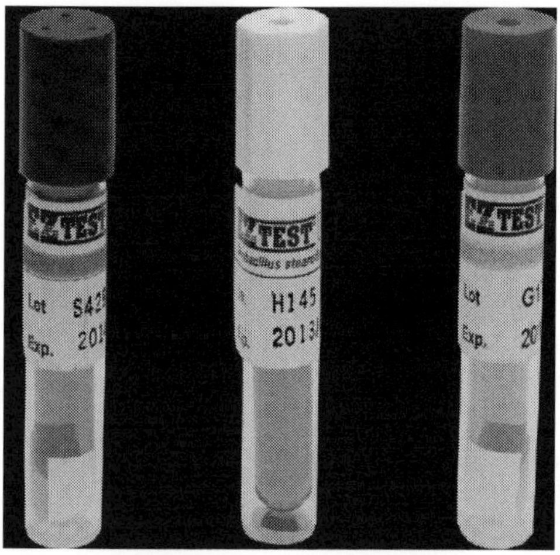

Figure 6.6: Biological Indicators.

https://biologicalindicators.mesalabs.com/wp-content/uploads/sites/31/2013/07/EZTestBlog.png

Cleaning

Cleaning is the elimination of distant substance, in general proficient by means of water by means of detergents or else enzymatic item for consumption. Cleaning is necessary sooner than elevated disinfection as well as sterilization for the reason that inorganic as well as organic substances so as to stay on the exterior of mechanism get in the way by means of the efficiency of these developments.

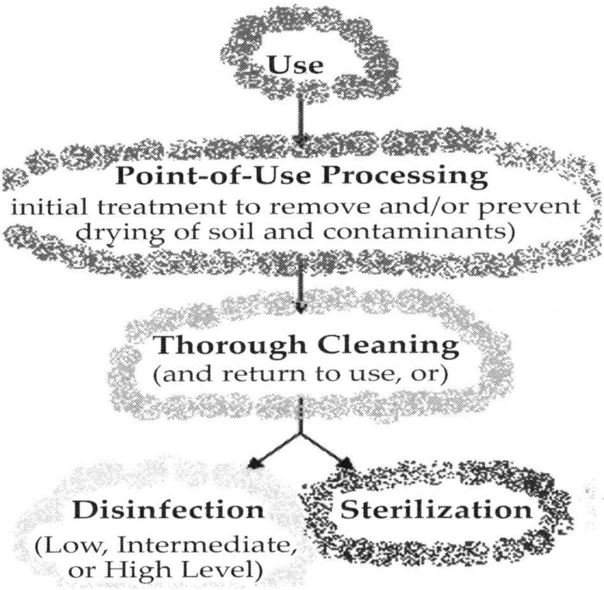

Figure 6.7: Cleaning and Disinfection in Food Processing.

https://jody929lee.files.wordpress.com/2015/02/ucm253015.png?w=336&h=440

Cleaning is done manually in use areas without mechanical units (e.g., ultrasonic cleaners or washer-disinfectors) or for fragile or difficult-to-clean instruments.

1) **Manually**

a) **Friction**: Old and dependable method, for example, rubbing/scrubbing the dirty region by means of a brush.

b) **Fluidics**: Used to get rid of soil as well as debris as of interior controls subsequent to brushing as well as, as the blueprint does not permit means of access of a meeting all the way through a medium, for example, fluids under pressure.

Figure 6.8: Clening by Fluidics.

https://encrypted-tbn0.gstatic.com/images?q=tbn:ANd9GcRSbmTBCcDx4Gf HiVpoQ_aEU3rRNLbJdKIXZdDH5ja1McmHN033

2) Mechanical or Automatic cleaners

a) **Ultrasonic Cleaners:** Get rid of top soil through cavitations as well as implosion by which waves of sound energy are propagated inside aqueous solutions on the way to interrupt the bonds so as to grasp particulate substance in the direction of exteriors. Users of ultrasonic cleaners ought to be conscious so as to the cleaning fluid might effect in endo-toxin infectivity of surgical apparatus, which might cause strict provocative responses.

Figure 6.9: Ultrasonic Cleaners.

https://www.pflegewelt.de/seniorenbedarf/food/Jeken/2014-Professionelle-und-Hot-Sale-Waschmaschine-Jeken-Dental-DIGITAL-Ultraschall-Reiniger-ps-30-a-6,5-l-geeignet-fuer-Labor,-Krankenhaus,-Elektronische-Werkstatt-von-Jeken-12095999.jpg

b) ***Washer- De-contaminators/Disinfectors***: Perform similar to a dishwasher so as to uses a mixture of water movement as well as detergents in the direction of removes top soil.

Figure 6.10: Washer de-contaminator.

http://img.medicalexpo.com/images_me/photo-m2/70928-6320457.jpg

c) ***Washer-Sterilizers***: Customized steam sterilizers so as to sparkling by means of satisfying the assembly room by way of water as well as detergent all the way through which steam overtakes on the way to supply agitation.

Figure 6.11: Washer-Sterilizers.

https://jody929lee.files.wordpress.com/2015/02/fows.jpg

In favor of apparatus clean-up, an impartial or else near-neutral pH detergent solution usually is used for the reason that such solutions usually offer the greatest substance compatibility outline as well as high-quality top soil elimination.

Disinfection

A lot of disinfectants are exploited unaccompanied or else in amalgamation within the healthcare surroundings. These comprise chlorine as well as chlorine compounds, alcohols, iodophors, formaldehyde, ortho-phthalaldehyde, glutaraldehyde, hydrogen peroxide, phenolics, peracetic acid, as well as quaternary ammonium compounds. The explanation of a few of these disinfectants is given.

1) Chemical Disinfectants

a) Alcohol

Figure 6.12: Alcohol Disinfectants.

http://www.ainsworthdental.com.au/catalogue/images/P/AIIA5.jpg

Overview

Alcohol refers to two water soluble chemical complex, isopropyl alcohol as well as ethyl alcohol, which have commonly underestimated germicidal distinctiveness. These alcohols are speedily bactericidal to a certain extent than bacteriostatic alongside vegetative forms of microorganisms. The mainly practicable enlightenment in favor of the antimicrobial accomplishment of alcohol is denaturation of proteins. Methyl alcohol, methanol has the weakest bactericidal exploit of the alcohols as well as, as a result seldom is used within healthcare.

Uses

Not suggested meant for disinfecting medicinal as well as surgical supplies chiefly for the reason so as to they be short of sporicidal accomplishment as well as they cannot break through protein-rich supplies. Alcohols have been used efficiently in the direction to sterilize oral as well as rectal stethoscopes, hospital papers, thermometers, and scissors.

b) Chlorine and Chlorine Compounds

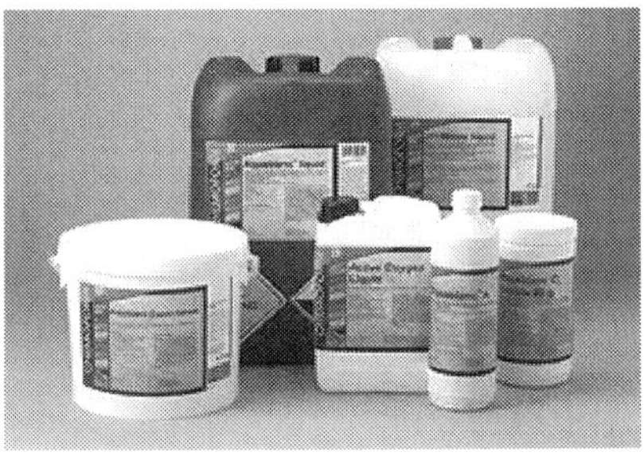

Figure 6.13: Chloine and Chlorine Compounds Disinfectants.

https://jody929lee.files.wordpress.com/2015/02/group-products.jpg

Overview

Hypochlorite, the most broadly used of the chlorine disinfectants, is accessible as liquid or else solid. Commonly known as the household peroxide. They contain a wide range of antimicrobial action, do not run off poisonous remains, are unaltered as a result of water hardness, are economical as well as speedy performing, get rid of dried out otherwise unchanging organisms as well as biofilms as of exteriors, as well as comprise a little occurrence of severe toxicity.

Uses

Hypochlorites are extensively exploited into healthcare amenities in a selection of situations. Non-living chlorine solution is used in favor of sterilizing tonometer heads as well as in favor of spot-disinfection of countertops as well as floors. Chlorine extensive has been used the same as the antiseptic in water action.

FACTORS AFFECTING THE EFFICIENCY OF DISINFECTION AND STERILIZATION

The action of germicides alongside microbes relies on an amount of aspects, a number of which are fundamental character of the individual, others of which are the substance as well as exterior substantial surroundings. Responsiveness of these issues ought to show the way on the way to improved bring into play of sterilization as well as disinfection methods as well as will be for a short time examination.

Factors counting

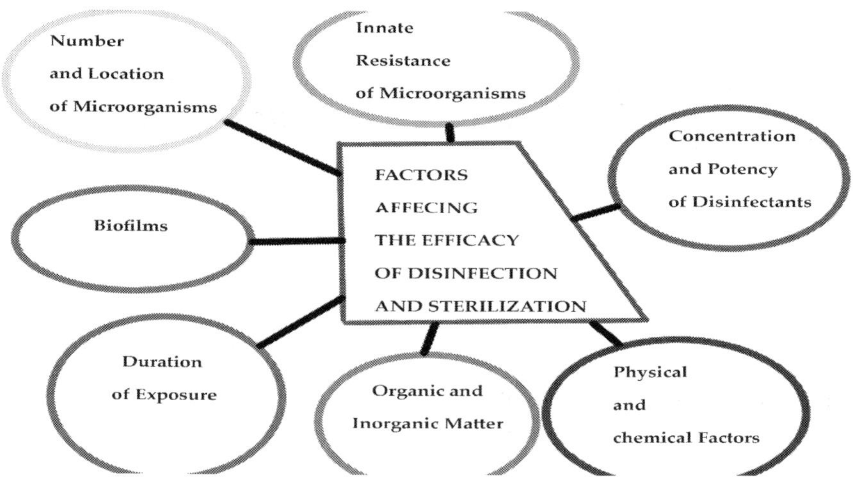

Figure 6.14: Factors Affecting the Efficiency of Disinfection and Sterlization.

https://jody929lee.files.wordpress.com/2015/02/factors-affecting-of-disinfection-and-sterilization.png?w=640&h=435&crop=1

1) Number and Location of Microorganisms

The superior the quantity of microorganisms, the additional instances disinfectant requirements in the direction of annihilate the entire of them. Researchers in addition have exposed so as to comprehensive or else bundle units are supplementary easier said than done on the way to inactivate than mono-dispersed units.

2) Innate Resistance of Microorganisms

Microbes fluctuate really inside their confrontation on the way to chemical sterilization as well as germicides methods. On the way to demolish the mainly opposed to categories of microbes, the consumer requirements on the way to make use of introduction times as well as a application of disinfectants desirable on the way to attain absolute devastation. Apart from in favor of prions, bacterial spores own the uppermost instinctive confrontation in the direction of substance germicides.

3) Concentration and Potency of Disinfectants

The additional strenuous the antiseptic, the superior its effectiveness as well as the shorter the occasion essential on the way to accomplish a microbial kill. Bearing in mind the extent of the disinfection occasion, that relies on the effectiveness of the disinfectant, in addition is significant.

4) Physical and Chemical Factors

a) Temperature

The movement of nearly all sanitizer increases because the temperature increases. Elevation in temperatures sources the sanitizer on the way to mortify as well as deteriorates its germicidal movement as well as therefore may manufacture a probable physical condition danger.

b) PH

Enhancement in pH recovers the antimicrobial action of a number of sanitizers. The pH manipulates the antimicrobial movement as a result of changing the sanitizer particles or else the sect exterior.

c) Relative Humidity

Relative humidity is the distinct mainly significant feature manipulates the movement of gaseous disinfectants as well as sterilants, like EtO, formaldehyde, as well as chlorine dioxide.

5) Organic and Inorganic Matter

Organic matter within the shape of serum, pus, blood, or else fecal otherwise lubricant substance knows how to obstruct by means of the antimicrobial movement of antiseptics within in any case two techniques. Non-living contaminants of microorganisms to the whole sterilization processes results, as of an occlusion in salt crystals. This supplementary give emphasis to the significance of meticulous cleaning of medicinal devices previous to any sterilization otherwise disinfection processes for the reason that mutually organic as well as inorganic soils are without problems uninvolved as a result of cleaning.

6) Duration of Exposure

Materials must be exposed on the way to the sanitizer in favor of the suitable least make contact with time. The precise times in favor of making germ-free therapeutic items are somewhat elusive because of the consequence of the aforesaid factors on disinfection effectiveness. Definite make contact with times have demonstrated dependable excluding, in general, longer contact times are additional efficient than shorter contact times.

7) Biofilms

Microorganisms might be sheltered as of disinfectants as a result of the manufacture of substantial masses of cells in addition to extracellular substances, or else biofilms. These are microbial communities so as to be powerfully attached on the way to the exteriors as well as cannot be effortlessly removed. On one occasion these masses structure, microorganisms surrounded by them know how to be opposed to disinfectants as a result of manifold methods, as well as substantial characteristics of grown-up biofilms, genotypic differentiation of the microorganisms, microbial manufacture of neutralizing enzymes, as well as physiologic incline inside the biofilm, for example, pH.

CHAPTER

7

FOOD PACKAGING MATERIALS

PACKAGING MATERIALS

Various different kinds of food exposed materials are commonly used every one with several possessions. It may also be an origin of chemical food contamination along with protection of foodstuffs by food packaging. "Food contact material" is the layer that is in direct contact with the provisions.

In many different forms packaging for foodstuffs appear, based on technical demands all over the supply sequence, as well as marketing wants similar to brand identity otherwise customer information and additional criterion.

Food contact material evaluates the name for some types of food wrapping; an artificial synthetic container is formed by plastic as well as has this material nature in exposure with the foodstuff. The foodstuff packaging is glass and encrusted metal from the closure for glass jars. The direct food contact coating is not carton, but coated plastic in the case of beverage cartons. For aluminium containers a coating is in direct exposure with the beverage. Definite kinds of paper may also be encrusted for illustration with a grease-proof covering

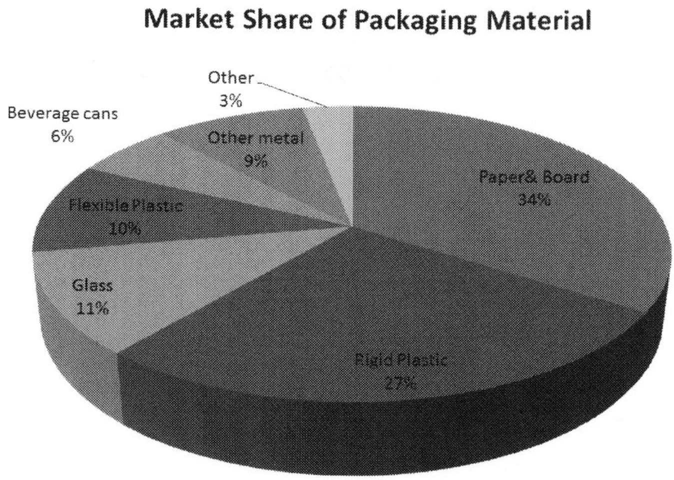

Figure 7.1: Food Packaging Materials.

http://www.foodpackagingforum.org/fpf-2016/wp-content/uploads/2014/06/FoodPackagingMaterials.png

The phrase food contact material utilized for provisions and beverage packaging, but as well to any other equipments that come into exposure with chow, while storage, processing as well as filling, or expenditure like as cooking paraphernalia.

In frequent, any food exposed material must not discharge chemicals into the victuals at levels that may damage human health. To assess the health impact related to food covering it is too essential therefore:

- Chemical composition of the wrapping substance
- Migration process that is the concentrations at which these compounds can partition into foodstuffs.

VARIOUS PACKAGING MATERIALS

Plastics

Plastics are organic substances made by macro cells called polymers. These polymers are large groups of monomers linked through a chemical process called polymerization. Plastics provide the necessary balance of attributes that cannot be achieved with other materials such as: color, lightweight, soft touch and resistance to environmental and biological degradation.

Properties and Characteristics of Plastic

The properties and characteristics of many plastics (though not always fulfilled in certain special plastics) are these:

- Easy to work and shape
- Have a low production cost.
- Possess low density.
- Tend to be waterproof.
- Good electrical insulators.
- Acceptable acoustic insulation.
- Good thermal insulation, but mostly cannot withstand very high temperatures.
- Resistant to corrosion and most of the chemical factors.

Some are not biodegradable or easily recyclable, and if they burn, are highly polluting.

Production Process of Plastic

The first part of the production of plastics is the production of polymers in the chemical industry. Today the recovery of post-consumer plastic is also necessary. Part of the industry-finished plastic directly used as grain or resin. More frequently, different forms of molding injection, compression, rotation, inflation, etc. or profile extrusion or yarns are used. Greater component of the plastic process is performed in a horneadora machine.

Classification of Plastics

This classification is considered the starting of the monomer from which part of the polymer production.

***Natural*:** These are polymers whose monomers are derived from natural products with specific characteristics, for example, cellulose, casein and rubber. In two of these examples there are other plastics which come:

Or cellulose derivatives are: the celluloid, cellophane and Cellon.

Or rubber derivatives include rubber and ebonite.

***Synthetic*:** Those that originate in man-made, mostly petroleum products such as polyethylene bags products.

According to their behavior in heat

***Thermoplastics*:** A thermoplastic is a plastic that, at room temperature, is plastic or deformable, it becomes a liquid when heated and hardens to a glassy form when cooled sufficiently.

Plastic Recycling

Severe environmental problems may be posed by the gradually rising manufacture of plastic, which include consumption of fossil fuels as well as cumulation of plastic dissipate in landfills as well natural environments high energy requirements while production. In the perspective of packaging, looms to decrease otherwise slow-down the requirements for virgin plastic have been developed as well as are by now apply to a lot of extents. These developments contain the diminution (**r**eduction) of packaging weight and/or quantity, the **r**euse of packaging as well as the **r**ecycling of particular polymers (**3R**).

Recycling Steps

Identification and Sorting

In 1988, the Society of the Plastics Industry (SPI) devised the resin identification code (RIC) aiming at the proficient recognition as well as partition of different plastics. In 2013, ASTM International issued the substitute of the three "chasing arrows", which are frequently related with recycling, by a solid equilateral triangle sign to focus only on resin identification, not on recyclability.

RIC Code	Material	Symbol "Chasing arrows"	Symbol "Triangle"
1	PET	1 PETE	1 PETE
2	HDPE	2 HDPE	2 HDPE
3	PVC	3 V	3 V
4	LDPE	4 LDPE	4 LDPE
5	PP	5 PP	5 PP
6	PS	6 PS	6 PS
7	others	7 OTHER	7 OTHER

Table 7.1: Resin identification code, old and new symbols.

Primary Automatic Plastics Recycling

It explains the products conversion with equivalent properties by thermoplastic polymers. For plastics, such closed-loop procedures can merely be useful which have been completely cleaned as well as separated from polluted plastics before recycling at all or which have not been used.

Secondary Automatic Plastics Recycling

It usually leads to products of lower mechanical properties. Thermoplastics' minimization of the polymer's molecular weight and modifications in physico-chemical characteistics frequently have to be taken into account contain just one polymer (e.g. PE, PP, PS, PET, and PVC) may be automatically recycled after use.

Tertiary Chemical Plastics Recycling

It illustrates methods by which polymers are biologically otherwise chemically tarnished into smaller molecules which may serve as new building masses for more chemical syntheses.

Quarternary Plastics Recycling Procedure

Through incineration, it recovers energy from plastics.

Recycled Materials

Recycled Polyethylene Terephthalate (PET)

For water as well as soft-drinks worldwide PET has become the most frequently used packaging material. The recyclability as well as availability of gether PET bottles promoted the development of PET recycling processes as well as public concerns about environmental effects of PET disposal. In the beginning, in the production of polyester fibers, PET recyclates were mostly used.

Supposed "challenge tests" were established to measure whether a recycling process can decrease any contamination below a set limit as well as thus fulfill with the legal supplies. Further PET degradation products, process chemicals or food components can result in unwanted impurities.

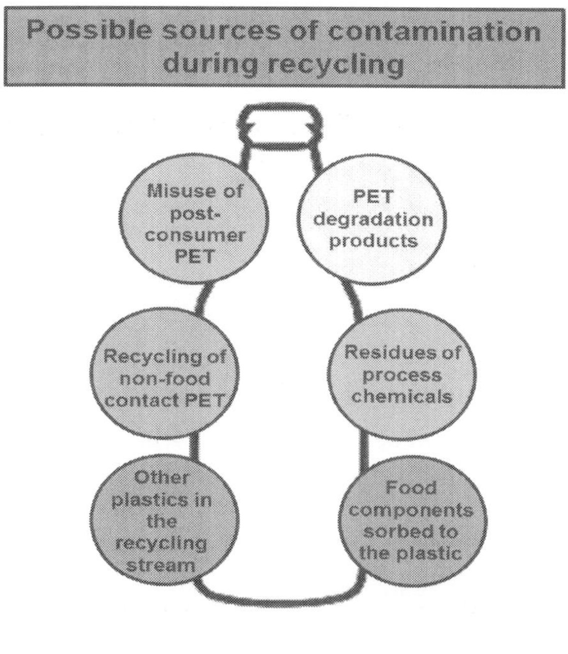

Figure 7.2: Sources of Contamination While Recycling.

In materials recovery facilities a characteristic PET recycling procedure begins with the collection, separation and sorting of post-consumer PET bottles. In recycling facilities post-consumer PET is washed to remove glue, dirt and food leftovers and then ground into chips.

CAN COATINGS

Epoxy-based can coatings are recently being replaced by alternatives because of toxicological evidence, public discussions, and current regulatory decisions.

Food and beverage cans preserve the taste and nutritional values of their filling for up to many years. As a result of such long storage times, the interactions between the packaging and the food need to be minimized. Cans are typically coated with an organic layer that protects the integrity of the can from effects of the food and prevents chemical reactions between the can's metal and the food. To fulfill the technical and legal demands, can coatings should withstand:

- The production and sterilization processes.
- Be universally applicable for all food and beverage kinds.
- Prevent chemical migration into food in quantities that endanger human health.
- Adhere to the can even after non-intentional deformation.
- Resist aggressive food kinds and protect the metal of the cans.
- Preserve the food.
- Maintain its organoleptic properties over a long duration.

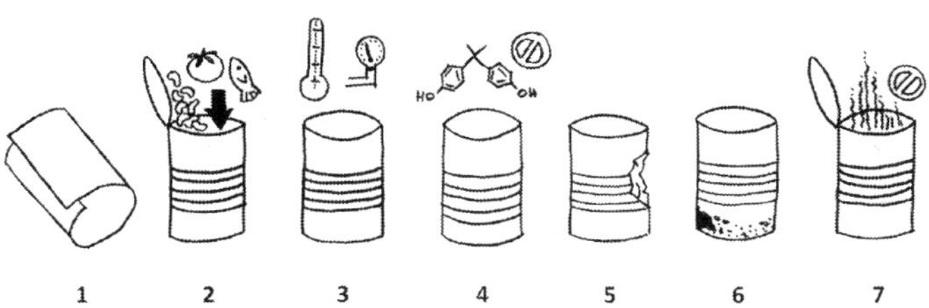

Figure 7.3: Can packaging.

http://www.nzdl.org/gsdl/collect/hdl/index/assoc/HASH0102.dir/fig3f22.png

Can Production and Market Data

Cans are made of three types of materials: aluminum, tin-coated steel (tinplate) and electrolytic chromium coated steel (ECCS). Can bodies are either formed as 3-piece welded cans (3PC), 2-piece drawn and redrawn (DRD) cans or as 2-piece drawn and ironed (D&I) cans. Independently of the material and the production process, most cans are coated internally and externally with films of 1 to 10 μm thickness. Coatings are commonly applied to both sides of planar metal sheets or coils by roller coating before the cans are formed. Alternatively, coatings are sprayed on preformed cans. Tin cans are used without internal coatings for light colored, acidic juices and fruits (e.g. pineapple, pears, peaches), because tin is more easily oxidized than the food, thus preventing darkening and flavor changes posed by oxidation of the fruits. Food cans are usually pressure-sterilized with the exact situation depending on the food type. However, beverage cans may be pasteurized or sterilized in the sealed cans or filled under aseptic situations.

Coatings: Properties and Alternatives

Many different can coatings are commercially available, but most of them are based on a limited number of chemicals functionalities. Coatings consists of different additives, e.g. agents to increase surface slipping as well as abrasion and scratch resistance of can coatings, lubricants, anti-foaming agents, adhesives, scavengers for hydrochloric acids, and pigments.

Epoxy-based coatings have the highest market share of more than 90%. However, can manufacturers and food companies have started to replace BPA-based epoxy coatings by alternatives as a consequence of toxicological results, public discussions, and recent regulatory decisions. Acrylic and polyester coatings are recently used as first generation alternatives to epoxy coatings and, more recently, polyolefin and non-BPA epoxy coatings were developed. More inventions include BPA capturing systems and top coatings. Most of these alternative coatings are more expensive than epoxy coatings and may not display the same array of characteristics with respect to their stability and universal applicability yet.

Epoxy Coatings

In the 1950s, epoxy resins were introduced as coatings for aluminum and steel cans. Their stability, protective function, and technical properties made them the most frequently used coating material. Most epoxy coatings are synthesized from bisphenol A (BPA, CAS 80-05-7) and epichlorohydrin forming bisphenol A-diglycidyl ether epoxy resins. Many different blends of epoxy coatings were developed with epoxy-phenolic coatings being the most significant subgroup. Other blended resins are e.g. epoxy amines, acrylates, and anhydrides.

Oleoresins

First can coatings were made of oleoresins, which are blend of oil and resin extracted from plants. Oleoresins are rather flexible and easily applied, but do not adhere well to metal surfaces, have a limited corrosion resistance, and need long curing times. Furthermore, they may change the organoleptic properties of food.

Vinyl

Vinyl coatings are produced from vinyl chloride and vinyl acetate. They are highly flexible and stable under acidic and alkaline conditions, but they do not adhere well on metal and do not withstand at high temperatures. Vinyl coatings need plasticizers and stabilizers and are mostly blended with other resins. Vinyl organosols are prepared from suspensions of resin in organic solvent. Organosols offer comparably higher chemical resistance, thermal stability, and adhesion properties than vinyl coatings.

Phenolic

Phenolic resins are composed of phenols and aldehydes. They are highly corrosion resistant and protect cans from sulfide staining. Phenolics have low flexibility, do not adhere with metal, and may change the odor and flavor of foods. They are applied as coatings for drums and pails, but un-blended phenolic resins are not used in food and beverage cans. However, phenolics are common cross linkers (e.g. in epoxide resins) and enhance their resistance.

Acrylic

Acrylic resins are most frequently synthesized from ethylacrylate. They have a clean appearance and display corrosion and sulfide stain resistance, but they are brittle and may change the taste and odor of foods.

Polyester

Isophthalic acid (IPA) and terephthalic acid (TPA) are the main carboxylic acids used in polyester coatings. Polyester resins are easy to handle during the production process and adhere well to the metal surface, but they are commonly not stable under acidic conditions and have a poor corrosion resistance. Alternatively, polyethylene terephthalate (PET) coatings are used to laminate beverage cans, but adhesives are needed to interact the PET onto the metal.

Polyolefins

Coatings that are based on dispersions of polyolefins have currently entered the market. According to the manufacturer, the final polyolefin coating exhibits corrosion protection, adhesion, and flexibility without impacting the flavor of the food.

PHTHALATES

For some phthalates food is probable to be the chief contact way, in addition to food contact materials are likely an appropriate cause. Phthalates are general plasticizers utilized in polyvinyl chloride (PVC), but migration from other categories of food packaging has also been originated. Their use is controlled in the EU although phthalates are legally used in food packaging.

Phthalates are additives commonly added to PVC for softening, therefore they are also known as plasticizers. However, they have also been found to migrate from non-PVC food contact materials. In the production of polypropylene plastics phthalate-based catalysts are also used. Introduction to phthalates is of concern, as these substances are associated to reduce fertility, reproductive toxicity as well as testicular toxicity in animal learnings. Increased levels of phthalates are associated with severe health effects in people, for example obesity as well as reduced masculinization in infant boys. The main exposure source, for definite phthalates are DEHP, DBP and DIBP in food. Since phthalates are lipophilic, increased phthalates concentrations in people were present to be associated with the consumption of specific food groups, particularly fatty foods.

The European Commission has limited the utilization of phthalates in food contact materials made of plastic, since July 2008. In a 2008/2009 control campaign different packaged food contact materials were sampled in Denmark as well as analyzed for the presence of phthalates. However, in a record study one year afterward no phthalates were found in closures, most phthalates were present to migrate from glass jar closures manufactured outside the EU. Other phthalate start was single-use plastic gloves for catering, conveyor belts, as well as plastic tubes for liquid foodstuffs like milk. Phthalate limits did not minimize as visibly in these other product groups in the follow-up sampling, in addition to non-compliant products might still be on the market.

From Italy, a 2011 scientific study establish high concentrations of phthalates DEHP, DBP in school lunches, sampled at some stage in 2010. Phthalate concentrations were shown to enhance after packaging and transport in non-PVC caniters composed of polyethylene-coated aluminium as well as polyethyleneterephthalate-coated aluminium. Highest concentrations of phthalates were present in bread, a finding that was currently confirmed for the Belgian market. Especially since bread is not a fatty food, it is unclear why phthalate concentrations were present to be higher in bread.

GLASS

Near 3000 BC the first glass substance intended for sharing provisions is supposed to have came out, tumbler has an enormously extended account in foodstuff wrapping. The manufacture of tumbler canisters occupy warming a blend of the glass former that is silica, the melting negotiator sodium carbonate, as well as limestone or calcium carbonate plus stabilizers to elevated warmth in anticipation of the source soften into a substantial fluid group that is subsequently transferred into shape, is alumina. Broken recycled tumbler is as well utilized in glass construction as well as might report in favor of to the extent that 60% of all unprocessed resources. Tumblers canisters apply in food covering be repeatedly surface-coated to supply lubrication in the invention streak as well as reduce scrape otherwise exterior scratch as well as contour jams. Glass varnishes as well improve and preserve the potency of the bottle to decrease fracture. Enhanced rupture resistance permits producers to utilize thinner flute, which minimizes weight furthermore is enhanced for removal along with carrying.

As it is unscented plus chemically immobile by means of almost the entire chow foodstuffs, goblet has numerous compensate intended for food-packaging claims. It is resistant to gas as well as billows, thus it preserves manufactured supplies novelty for an extended interlude of era devoid of spoil flavor or taste. The aptitude to endure elevated dispensation heat makes flute helpful in favor of thermal sterilization of equally high-acid plus low-acid sustenance. It is inflexible offers fine filling, as well as may be formed in frequent shapes plus natures. The precision of tumbler permits customers to observe the invention; so far deviations in tumbler shade may be able to shield light-sensitive stuffing. At last, glass covering reimbursement the surroundings as it is recyclable as well as reusable.

Similar to any stuff, glass has a few shortcomings. In spite of hard work to utilize thinner tumbler, its intense mass inserted to shipping outlay. A further disquiet is its fragility as well as vulnerability to fracture as of inside demands, blow, otherwise thermal distress.

METAL

For all packaging forms metal is the most versatile. It presents a mixture of outstanding substantial shield as well as obstacle possessions, formability and ornamental prospective, recyclability, as well as customer reception. In packaging 2 metals aluminum and steel are used most chiefly.

Aluminium: Aluminum is a silvery white, lightweight metal present in bauxite ore, wherever it may found in mixture by means of oxygen like alumina, as well as normally utilized to make foils, cans, as well as plastic otherwise laminated paper packaging. To improve the strength properties of aluminium, magnesium and manganese are frequently added.

It is moreover a perfect substance in favor of reprocessing as it is simple to recover as well as procedure into novel foodstuffs. Uncontaminated aluminum is utilized in favor of light covering of mainly pet food, soft-drink cans, pretreated closures as well as seafood. The mainly recurrent shortcomings of aluminum are its elevated price in contrast to other metals for illustration, steel.

Aluminum Foil: Aluminum foil is made by rolling pure aluminum metal into very thin sheets, followed by annealing to achieve dead-folding properties, which allows it to be folded tightly. As the entire aluminum coverings, foil supplies an outstanding hurdle to air, moisture, light, odors as well as microbes. It is inert to acidic foodstuff as well as doesn't need lacquer otherwise supplementary shield. Though aluminum is recyclable devoid of difficulty, foils cannot be prepared as of reused aluminum exclusive of pinhole development in the slight pieces.

Laminates and Metallized Films: Lamination of covering has the requisite of aluminum foil to paper otherwise artificial coat to develop obstacle charecteristics. Thin estimates assist appliance. Even though, lamination to synthetic capables warmth sale capacity, the seal doesn't thoroughly bar moisture as well as air. Metallized foils are synthetics having a thin layer of metal, aluminum. Even if, the single constituents of metallized filmsas well as laminates are strictly recyclable, the complexity in separating as well as categorization of the substance prohibits cost-effectively possible reprocessing.

Tinplate: formed as of low-carbon brace, so as to be, black plate, tinplate is the effect of covering equally both surfaces of black plate by thin coatings of tin.

Because it has good ductility and formability, tinplate can be used for containers of many different forms and shapes. Tinplate is an outstanding substrate designed for current metal covering as well as lithoprinting machinery, making possible exceptional graphical ornamentation. In conclusion, tinplate is without difficulty used many times devoid of defeat of worth in addition to is appreciably lesser in rate than aluminum.

Tin-Free Steel: Is also known as electrolytic chromium or chrome oxide coated steel. Even though the chrome/chrome oxide makes tin-free steel unsuitable for welding, this property makes it excellent for adhesion of coatings such as paints, lacquers, and inks.

PAPER AND PAPERBOARD

In later part of the 19th century, the utiliztion of paper as well as paperboards in favor of foodstuff covering dates reverse to the 17th century by means of speed up practice. The fibers are then pulped and/or bleached and treated with chemicals such as slimicides and strengthening agents to produce the paper product. Paper as well as paperboards is normally used in milk cartons, corrugated boxes, bags and sacks, folding cartons as well as packaging paper. Paper plates, cups, as well as tissue paper are other instances of paper as well as paperboard foodstuffs.

Paper

Simple paper may not be consumed to shield foodstuff in favor of lengthy time periods as it has underprivileged obstacle charecteristics as well as is not high temperature sealable. The many different types of paper used in food packaging are as follows:

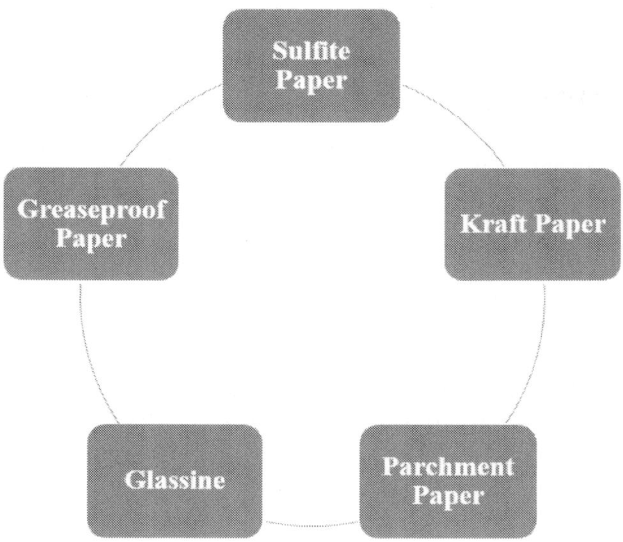

Figure 7.4: Types of Paper.

- **Kraft Paper**—formed through a sulfate management method, it is accessible at numerous appearances: unbleached, natural brown, bleached white as well as heavy duty.

- **Sulfite Paper**— Weaker as well as lighter than kraft paper, sulfite paper be fixed to develop its manifestation as well as to improve its soaked potency as well as oil resistance. It is utilized to compose small wrappers otherwise bags for covering confectionary as well as biscuits.
- **Greaseproof Paper**— This paper is prepared by a method called beating, by which the cellulose fibers go through a longer than standard hydration interlude so as to causes the strands to divide as well as develop into viscous. Greaseproof paper is consumed to enclose cookies, snack foods, other oily foods, as well as candy bars, an exploit so as to be being replaced by synthetic films.
- **Glassine**—Glassine is greaseproof paper in use to an excessive additional hydration to make an extremely opaque sheet by means of a shiny finish as well as extremely flat. It is cosumed like a facing for cooking fats, biscuits, baked goods, in addition to fast foods.
- **Parchment Paper**—Parchment paper is prepared as of acid-treated tissue passed by a sulfuric acid bath. The acid adjusts the fiber to compose it impervious as well as smoother to oil as well as water that adds quite a few soaked potency.

Paperboard

Paperboard is thicker than paper by means of an elevated weight per unit area as well as frequently prepared in manifold coatings. It is normally utilized to compose cans for shipping, like cartons, boxes, as well as trays, in addition to infrequently utilized meant for direct foodstuff exodus. A variety of categories of paperboard are as subsequent:

- **White Board**— It is prepared by numerous thin coatings of lightened substance tissue, white board is normally consumed like the inside coating of a container.

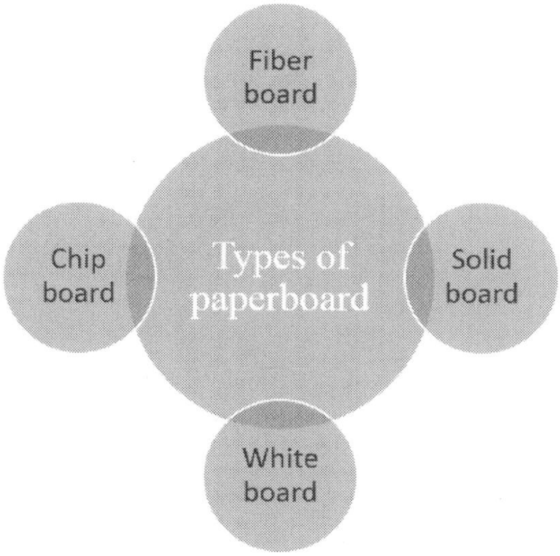

Figure 7.5: Types of paperboard.

- **Solid Board**— Having durabilityas well as strength, solid board has manifold coatings of blanched sulfate board. It is moreover utilized to enclose soft drinks as well as fruit juices.
- **Chipboard**—Chipboard is prepared as of reused paper as well as frequently has blemishes as well as a contamination as of the original paper that pepares it, is inappropriate for undeviating exposure with foodstuff, folding, as well as printing.
- **Fiberboard**—Fiberboard may be corrugated otherwise solid. The solid category has a central white board coating as well as external kraft coating as well as supplies good quality shield against compression as well as impact. Fiberboard's resistance to crushing damage in addition to impact abrasion make it vastly utilized for shipping bulk foodstuff as well as case packing of retail foodstuffs.

Paper Laminates

Paper laminates are uncoated otherwise coated papers based on sulfite as well as kraft pulp. They might be coated with aluminum otherwise plastic to develop different charecteristics. For illustration, paper be able to be plastic-coated by polyethylene to able it heat sealable as well as to develop moisture as well as gas, barrier charecteristics. Laminated paper is consumed to wrap up dehydrated foodstuff like herbs, spices, as well as soups.

UV FILTERS IN FOOD PACKAGING

UV light describes the part of the light spectrum between 10nm and 400nm. As such, UV light has a relatively high energy level, which may be harmful to materials. Some UV filters act as UV screens absorbing the harmful UV light. They are opaque and protect both food as well as the packaging material. Other UV screens are based on organic compounds and act as ultraviolet degradation inhibitors.

The following UV screens have been measured in food stuffs and are formed in quantities larger 1000 tonnes/year:

In polymers:

- 2-(2H-benzotriazol-2-yl)-4-methyl-phenol (CAS 2440-22-4) showed estrogenic activity in Yeast Two-Hybrid Assay and in the Estradiol receptor ELISA assay
- Bumetriziole (CAS 3896-11-5) showed estrogenic activity in Yeast Two-Hybrid assay, a slight increase in the proportion of fetuses with
- incomplete ossification of sternebrae at a dose level of 3000 mg/kg bw/day in mice (repro tox study, OECD, report 2010), irritating to the eye, contact sensitizer (New Zealand EPA)
- 2,4-bis(1,1-dimethylethyl)-phenol, phosphite (3:1) (CAS 31570-04-4) showed estrogenic activity in Yeast Two-Hybrid assay.

BIOPLASTICS

Types and Application of Bioplastics

As an alternative to the conventional fossil fuel based plastics, bioplastics are increasingly being used in food contact materials (FCMs). For example, the Coca Cola Company has recently initiated its Plant Bottle, which is partially made from biobased plastics and Danone is using polyactide (PLA) for its yoghurt cups. Two different kinds of bioplastics exist: biobased polymers and biodegradable plastics. Two further typical definitions for biorelated plastics are oxo-biodegradable plastics and bio-nanocomposites.

Biobased Polymers are made from biobased resources though in practice biobased resource amount may vary. Biomass used for the production of bioplastics may either be extracted directly from plants (starch, cellulose) or formed by microorganisms in fermentative processes (e.g. polyhydroxyalkanoates (PHA)).

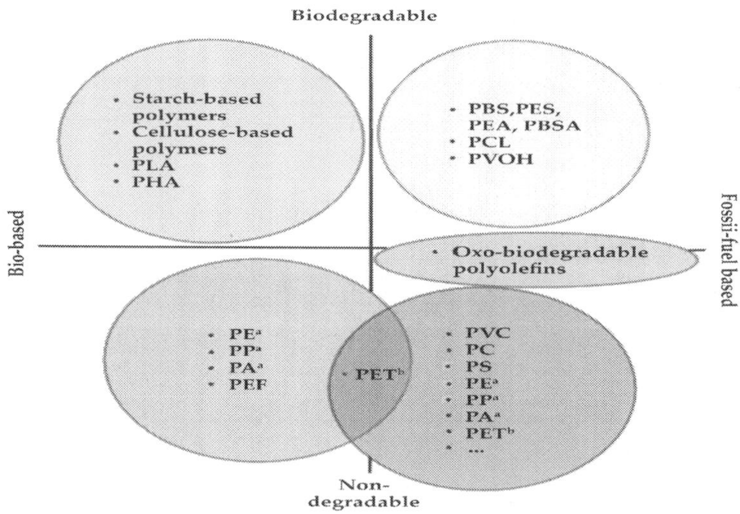

Figure 7.6: Bioplastics – Material source and biodegradability.

http://www.foodpackagingforum.org/fpf-2016/wp-content/uploads/2014/06/Figure-1-800x600.png

Biodegradable plastics may be made from both natural and fossil resources and are biodegraded by microorganisms in their natural environment.

Oxo-Biodegradable Plastics are mostly composed of polyolefins such as polyethylene (PE) and polypropylene (PP), which consists of further chemical additives, intended to enhance degradation.

Bio-Nanocomposites are biopolymers which have been stabilized using nanoparticles. The nanoparticles enhance technical charecteristics, such as barrier, thermal, chemical or mechanical stability and include nanoclays and nanosilver.

Toxicity

The monomers of cellulose- and starch based polymers as well as those of PHB and PLA are judged to be of no health concern. This stands in contrast to the toxicological characteristics of many other monomers used in plastics packaging materials (e.g. bisphenol A (BPA), bisphenol S (BPS), vinyl chloride and acrylamide). However, many additives are usually used in bioplastics than in conventional plastics. The several physicochemical properties of biobased FCMs may result in higher or lower migration rates of additives.

Market, Regulation and Standards

Bioplastics covers around 1% of the global plastics demand according to the European Bioplastics association and Plastics Europe. In 2007, starch had 50% of the market share of bioplastics, followed closely by PLAs.

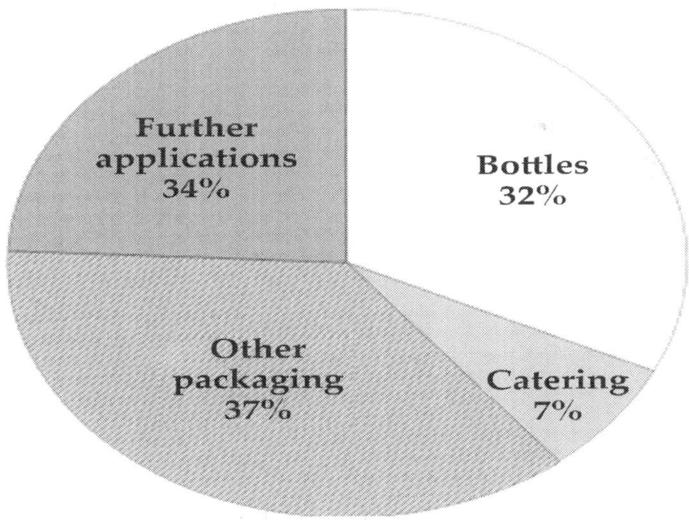

Figure 7.7: Percentages of materials used for packagings.

http://www.foodpackagingforum.org/fpf-2016/wp-content/uploads/2014/06/Figure-2-400x300.png

While currently only 0.01% of arable land mass is used for cultivation of raw materials used for bioplastics, the substitution of all plastics with bioplastics would require the use of 7% of the globally arable land. Plastics Europe and European Bioplastics suggest the use of sustainability certification schemes to ensure sustainable sourcing.

ROLES OF FOOD PACKAGING

The principal roles of food packaging are to protect food products from outside influences and damage, to contain the food, and to provide consumers with ingredient and nutritional information.

Protection/Preservation

Food packaging can retard product deterioration, retain the beneficial effects of processing, extend shelf-life, and maintain or enhance the quality and safety of food. In doing so, packaging provides protection from 3 main classes of external influences: chemical, biological, and physical.

Chemical Protection minimizes compositional changes triggered by environmental influences such as exposure to gases (typically oxygen), moisture (gain or loss), or light (visible, infrared, or ultraviolet). Many different packaging materials can provide a chemical barrier. For instance, plastic ceilings have some permeability to vapors as well as gases, the same as perform the gasket substances utilized in tops to help closure as well as in metal be able to cover to permit fastening after satisfying.

Biological Protection provides a fence to microbes, spoiling agents as well as pathogens, rodents, insects, as well as other natural world, thus prevent spoilage as well as disease. These obstacles utility by means of mechanisms multiplicity, counting preventing odor transmission, avoiding contact to the foodstuff, as well as sustaining the inside surroundings of the wrapping.

Physical Protection protects foodstuff as of perfunctory harm as well as have militates against the distress as well as trembling stumble upon while sharing. Suitable physical wrapping as well shields customers as of damaging vulnerabilities. For illustration, child-resistant closure hinders contact to potentially hazardous foodstuffs. Additionally, the replacement of plastic wrapping meant for foodstuffs assortment as of shampoo to soda bottles has reduced the threat as of wrecked glass cans.

Containment and Food Waste Reduction

Any appraisal of foodstuff packaging's contact on the surroundings ought to show the helpful reimbursement of reduced foodstuff waste is all through the contribution sequence. Considerable foodstuff consumption has been reported in several states, variety as of 25% in favor of foodstuff granules to 50% for vegetables as well as fruits. Insufficient protection or preservation, storage, as well as transport have been mentioned while poses of foodstuff dissipate.

Convenience

Convenience charecteristics like handling, ease of access, as well as product visibility, disposal; microwave ability as well as reseal ability significantly manipulate wrap up improvement. As a result, wrapping plays a fundamental function in reducing the endeavor essential to get ready as well as supply foodstuff. For instance, a cookie producer presently introduces a stretchy container with an achieved segment so as to supplies contact to the cookies. A covering by means of a peel able seal wraps the opening prior to sale as well as allows reclosure after opening. These handiness charecteristics insert worth as well as aggressive compensations to foodstuffs excluding might in addition control the concentration as well as category of wrapping dissipate necessitate removal.

Tamper Indication

Obstinate interfering with pharmaceutical as well as food products has resulted in particular wrapping charecteristics considered to eliminate otherwise minimize the hazard of adulteration as well as tampering. Though whichever wrap up be able to be breeched, tamper-evident properties cannot simply be substituted. Particular printings moreover have holograms so as to can't be effortlessly copied. An instance of a tamper-evident characteristic so as to needs no extra wrapping resources is a heat seal consumed on medicinal wrapping so as to be chemically prepared to modify color when unlocked.

Other Functions

Wrapping might provide other purposes, like a transporter in favor of qualities for illustration, addition of an award, extra creation, otherwise containers or else coupon for domestic utilization. The prospective for packaging use or reuse eradicates otherwise postponements access to the dissipate flow.

Source Reduction

Source diminution that is, waste anticipation is reducing the amount and/or toxicity of the waste ultimately generated by changing the design, manufacture, purchase, or use of the original materials and products. Some specific ways to achieve the source reduction are subsequent:

Light Weighting

Light weighting is the way to achieve source reduction and has been attained in the paperboard business by means of thinner estimation resources. For instance, Anheuser-Busch kept 7.5 million pounds of paperboard by means of minimizing the depth of its 12-pack container wrapping.

Reusable and Refillable Containers

One more method to attain source reduction is in the course of reuse. For instance, several glass containers, particularly bottles, are commonly recycled later than rinsing with influential detergents. This is partially as transporting; cleaning as well as collecting such cans presents logistical difficulties so as to guide to developed preferences in favor of 1-way cans.

COMBUSTION/INCINERATION

Combustion is the restricted smoldering of desecrate in a nominated capability, as well as is an progressively more striking substitute for fritter away so as to can't be composted otherwise recycled. Reducing MSW level through 70% to 90%, burning incinerators might be prepared to make condensation so as to know how to moreover supply generates electricity otherwise heat. Actually, synthetics are derivative as of fuel feed stocks as well as have an elevated warmth concentration, which is beneficial for waste-to-energy burning.

Mass-Burn Incinerators

Mass-burn incinerators allow all kinds of in its present condition MSW apart from meant for substances so as to are much large to set out by the supply classification. Incorporated dissipate is sited on a exasperate so as to go by the combustor even as air is compulsory for the scheme above as well as below aggravate to encourage full incineration. Mass-burn incinerators are dissimilar as of other MWCs as they burn up dissipate in a particular fixed compartment as well as are normally built on place.

Refuse-Derived Fuel Incinerators

Refuse-derived fuel (RDF) incinerators use waste that has been preprocessed to remove non combustibles and recyclables.

Modular Combustors

Like mass-burn incinerators, modular combustors acknowledge the entire dissipate devoid of preprocessing except are unsurprisingly slighter than mass smolder. Modular Combustors are typically manufactured off location as well as be able to be swiftly accumulated everywhere they are required.

Land Filling

Landfills supplies environmentally special discarding of the residual MSW as well as the remains of reprocessing as well as burning processes. The operations as well as location of landfills are managed by state as well as federal legislations, as well as today's landfills are cautiously premeditated arrangements in which dissipate is secluded as of the adjacent groundwater as well as environment. Emerged otherwise having passed through as of landfill fritter away, leachate includes suspended, soluble, otherwise miscible supplies as of dissipate. EPA is examining an adjustment in landfill blueprint called a bioreactor so as to might increase anaerobic and/or aerobic deprivation of organic waste as well as leachate.

Other Disposal Methods

Anaerobic Degradation

The major variety of deprivation that takes plce in landfills is anaerobic digestion otherwise degradation. In anaerobic digestion or degradation, microbes gradually break solid dissipate, chiefly organic-based resources for instance paper as well as wood, in the deficiency of oxygen into principally ammonia, carbon dioxide, as well as methane. To avoid groundwater contagion, leachate ought to be controlled in an arrangement, generally a mixture of storage systems as well as liners. Eventually, leachate is practiced by means of a management capability to build a secure deposit so as to might be predisposed of carefully.

Litter

Littering is the inappropriate removal of dissipate. As brew canisters are a perceptible constituent in garbage, 11 states have ratified bottle bills to assist make sure an elevated speed of reuse otherwise recycling as well as to decrease waste. These container otherwise bottle bills placed regulations command a lowest refundable drop on soft drink, beer, as well as supplementry beverage containers, thus as long as an financial inducement to make sure the revisit of utilized bottles. Drink cans made from plastics, metal, as well as glass have been the mainly remarkable recycling achievements as they are simply noticeable as well as prepared of single supplies so as to be ecological. On the other hand, recyclable wrapping might gradually assist get rid of nastiness with the exposures to marine as well as animal life posed by waste.

Current Disposal Statistics

The reduction in dissipate production is recognized in measurement to the reduction in entity's dissipate production charge. EPA investigates MSW by means of 2 processes:

(1) By supplies paper as well as paperboards, metals, glass, rubber, plastics as well as textiles, leather, food scraps, yard trimmings along with wood.

(2) By main manufactured goods category packages as well as containers, durable goods, nondurable goods, as well as

additional dissipates.

In the manufactured goods category, packaging as well as containers characterizes mostly dissipate as of foodstuff wrapping for instance milk cartons, soft drink cans, as well as boxes of cardboard. Long-lasting supplies comprise furniture in addition to furnishings, domestic appliances, rubber tires, and rugs along with carpets, electronics, in addition to batteries.

Limitations of Solid Waste Management Practices

Suitable waste management needs cautious financing, planning, transportation, as well as transportation. By population increase as well as financial expansion, solid waste production extends that causes various confronts.

Source Reduction Compared to Convenience

Conveniences as well as source reduction are mostly differing forces within foodstuff wrapping. Convenience charecteristics for instance dispensability, unit packages, in addition to microwave capability typically need supplementary wrapping that is openly at probability with resource reduction hard work. Likewise, tamper indication possessions in addition add to the concentration of waste production. Users utter what is formed by means of what they decide to procure, as well as business will manufacture what customers claim, stipulation might be made advantageously.

Two challenging trends, effects source reduction of wrapping substances. One development is in the direction of additional inexpensive volume collections so as to necessitate a smaller amount wrapping substance per unit of item for consumption. If the proportion of parcel proportions leftovers unvarying, enhanced range will exend the attachment magnitudes as a square purpose and increases the quantity at the same time as a cube purpose. Consequently, the volume enhances supplementary quickly, consequential in a smaller amount of wrapping per unit capacity. The drift in the direction of superior masses, as is characteristic in store clubs, consequently shows a resource reduction. The opposing tendency is for handiness as well as segment platefuls, into that entity segments are wrapped, thus raising wrapping convention.

If the entire of the foodstuff is used, element casing will enhance MSW. Though, bulky section masses meant for small families are able to lead to foodstuff waste, food turn out to be intolerable via chemical, biological otherwise physical, resources as well as, as a result boost entirety damages.

An unintentional unhelpful result of bottle bills is the entrance of potentially unhygienic substances keen on foodstuff surroundings while the brew cans are take in for deliverance. For illustration, stipulation, a bottle were utilized in favor of gasoline transfer, garden chemicals, otherwise any other nonfood utilize before go back, this contagion might cause a danger at the position of revisit if it were a foodstuff organization. Additionally, if the bottle were not sufficiently sparkling prior to reprocessing, the contagion might eventually shift to a novel wrap up prepared by means of the used supplies. These potential harms are able to be determined, excluding the expenses take away as of the comprehend compensation. The utiliztion of reprocessing inside as a substitute of foodstuff organization diminishes these apprehensions.

Land Filling Compared to the Environment

As landfills have the ability to pollute groundwater as well as air, construction, proper design, as well as organization are important to avert ecological harm. Before 1970, landfills were positioned on the easily available, least expensive, convenient lands; for instance marshes, wetlands, spent mines, gravel pits, as well as quarries.

Landfill gas secretions in addition to emmisions have carbon dioxide, methane, as well as more than 100 unlike nonmethane organic substances for instance toluene, benzene, moreover vinyl chloride. Air release principles involve gas gathering as well as management arrangement; additionally, arrangement with the intention of, integrates energy recuperation are buoyant.

Communal resistance to siting of landfills as well as incinerators meant for waste removal is illustrated by means of the acronyms 'NIMED, not in my election district', 'NIMTO, not in my term of office', as well as'NIMBY, not in my backyard'. The siting difficulty is consequently not merely a concern of technological implication however as well social, political, as well as economic. Effectual community taking part is an important constituent of a wide-ranging siting approach.

Combustion Compared to the Environment

Though, with the exemption of modular combustors, incinerators need significant original assets, as well as building acquires 3 to 5 years. Additionally, incineration results in air discharge so as to ought to be controlled as well as considered. A greenhouse gas, carbon dioxide, is unconfined whilst products resulting as of remnant energy, for example plastics are smolder. Pollution apprehensions contain the discharge of particulate substances; acidic gases predominantly nitrogen oxides as well as sulfur dioxide, halogens, dioxins, heavy metals, as well as products of imperfect burning. Halogens as well as dioxins are unrestricted commencing burning of chlorinated polymers, the majority of that is PVC, comprising in the order of 1%, MSW.

Cadmium as well as lead based additives for colorants as well as synthetics added to the heavy metal substance of MWC residue.Even though utilized in small concentrations; these metals contemplate in the residue like the polymers are smoldered off.

Considerations for Use of Different Packaging Materials

Some of these factors are interrelated: for example, the type of food and the properties of the packaging material determine the nature of food–package interactions during storage. Other times, the factors are at odds with each other: for example, single-serving packaging meets consumer needs, but the bulk packaging is better for environmental reasons.

Environmental Characteristics

As a comprehensive analysis of the substance as of fabrication to removal, life cycle investigation is significant in formative the ecological effect of a wrapping. The investigations incorporate a quantitative estimation of ecological costs, taking into account concerns for example energy consumption, waste generation, as well as material use. The sustainability goal inherent within the cradle-to-cradle concept imposing zero impact on future generations builds on life cycle analysis to address material and energy recovery as well. Additionally, novel wrapping supplies are being urbanized to make possible the aspiration of accurate maintanance.

Balancing Priorities

Preferably, a foodstuff parcel might comprise of resources so as to sustain the safety as well as quality of the foodstuff ad infinitum with no deprivation over time; are convenient, easy to use, in addition to attractive whilst passing on the entire applicable data; are prepared as of renewable supplies, producing no dissipate meant for elimination; in addition to are economical. Hardly still, if constantly, perform today's foodstuff correspondences get together this condescending objective.

In due course, the customer plays an important character in wrap up proposes. Purchaser needs make manufactured goods transaction, as well as the wrap up is an important sales instrument.

LEACHING PROCESS

During leaching process, the constituents of a solid material are released into a contacting liquid phase. Although some species may be more of an environmental concern than others, the leaching process is indiscriminant such that all constituents, e.g., major or minor matrix components as well as inorganic, organic and radionuclide contaminants, are released under a common set of chemical phenomena which may include mineral dissolution, desorption and complexation, and mass transport processes.

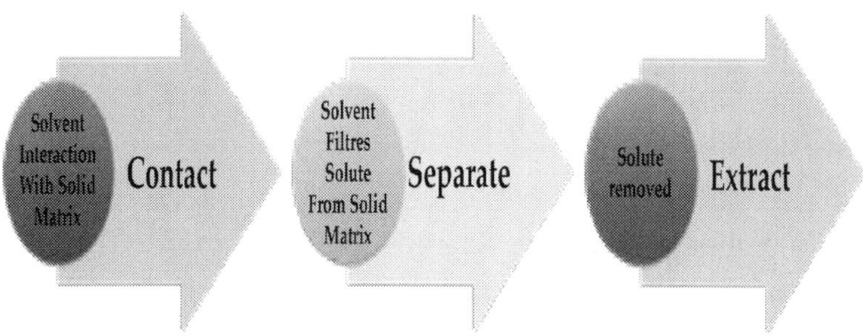

Figure 7.8: Step-by-Step Leaching Process for Chemical Industry Use.

http://study.com/cimages/multimages/16/leachingprocess.png

There are various specific factors that affets the rateand extnt of leaching. Among these factors are:

- Internal reactions (Physican and chemical)
- Envorinmental stresses
- erosion or cracking that cause physical degradation of the solid matrix
- Loss of matrix constituents due to the leaching process itself.

Factors Influencing Leaching

The summation of diffusion, hindered diffusion, tortuosity effects, and effective surface area effects through the pore structure of the material to the environment described as Mass transport is an important factor that influences leaching.

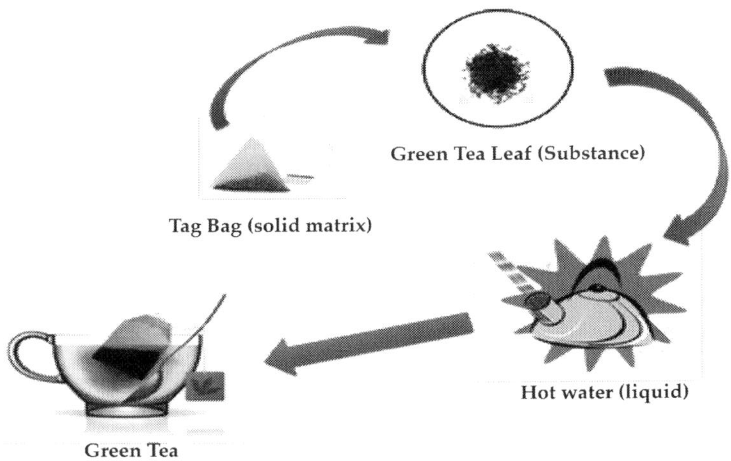

Figure 7.9: Example of Making Green Tea To Illustrate Leaching.

https://image.slidesharecdn.com/leaching-process1-160217192835/95/leaching-process-solidliquid-extraction-5-638.jpg?cb=1455737523

Solution pH, redox, presence of dissolved organic matter, and biological activity are important chemical factors that influence the liquid-solid partitioning (LSP) of a matrix constituent during the leaching process.

Physical factors (relative hydraulic conductivity, porosity and fill geometry) play a significant role in determining the rate at which constituents transport through a solid into a passing liquid phase.

The process itself is universal, as any material exposed to contact with water will leach components from its surface or its interior depending on the porosity of the material considered.

Next, the solvent travels through the solid matrix, separating the substance, or solute, from this matrix so that it can be collected. This step is commonly referred to as **percolation**, which is really just a fancy word the filter.

CONCLUSION

The primary purpose of food packaging must continue to be maintaining the safety, wholesomeness, and quality of food. However, the knowledgeable efforts by industry, government, and consumers will promote continued improvement, and an understanding of the functional characteristics of packaging. This would help to prevent much of the well-intentioned but ill-advised solutions that do not adequately account for both pre consumer and postconsumer packaging factors.

CHAPTER

8

FOOD INDUSTRIES

Food Industries in which chemicals are used for different functions, for better yield at low cost, few of them are explained below.

HARMFUL CHEMICALS PRESENT IN FRUIT JUICE DRINKS

A number of fruit juices and squashes have been found to contain a potentially lethal chemical, according to scientists. A worrying new study has found that fruit juices and squashes enjoyed by millions could have too much antimony – a potentially lethal substance when consumed in large doses.

Scientists have found high concentration of the chemical associated to **cancer** in a number of popular brands. In some cases, the limits of antimony were 10 times higher than what is considered 'safe' under EU guidelines. It is believed the substance is leaking into fruit **juice** from the plastic bottles, researchers believe. Specific brands have not been named, but it is believed that the chemicals were identified in 16 of the most popular cordials and fruit drinks enjoyed by children.

The team who carried out the study are concerned by their findings, published by the Royal Society of Chemistry, and have called for an immediate investigation. Claus Hansen, a PhD student at the department of pharmacology, who took part in the study, suspected that the citric acid present in the juices could speed up the leaching process.

'The **Antimony** concentrations in the products tested exceeds the limit of drinking water but no legislation exists for foodstuffs, so no legislation has been broken. 'However, we cannot be sure that the antimony limits are harmless. It would be a good idea to have some more research to get a better impression of what the antimony limit should be in fruit juices.'

Figure 8.1: Fruit juice.

http://www.healthydietbase.com/wp-content/uploads/2014/04/bigstock-fresh-fruit-juices-on-wooden-t-54817562-840x560.jpg

In 2005, **Volvic** mineral water was at the heart of a health scare after a potentially harmful substance was found in bottles. An investigation found that the water contained naphthalene – a chemical that can pose liver damage if drunk in high quantities.

BREAD

Staple food in many countries to fulfill their hunger, but now it has many harmful chemical residuals that may cause much harm to the consumer.

CHEMICALS IN VEGETABLE GROWTH, FRUIT RIPENING AND HEALTH EFFECTS

With the world's population reaching over seven billion in the 21st century and concerns over climate variation and global warming taking centre stage, the need has equally arisen over how to feed the worlds increasing population. This has resulted in scientists and agriculturists introducing what have come to known as genetically modified foods. This has more resulted in concerns over these genetically modified foods, and fruit ripening chemicals and their health effects.

Asif, M. (2012), lists some of the most frequent chemicals used in fruit ripening and growth as, "...calcium carbide, acetylene, ethylene, propylene, ethrel (2-chloroethyl phosphonic acid), glycol, ethanol..." Asif further contends that, "calcium carbide (CaC_2) is the generally used chemical for ripening of fruits, due to its low price and availability in local market; however, use of this chemical in fruit industry is being discouraged worldwide due to dangers of explosion and carryover of toxic materials like arsenic and phosphorus to consumers, thus making the healthy fruit poisonous"

Thus the issue of the health effects of these chemicals arises. Asif postulates that, this calcium carbide has high traces of 'heavy metal arsenic and phosphorous' and thus, "People exposed to high concentrations of arsenic can have nausea and vomiting, diarrhea, anemia, and low blood pressure. These signs may be followed by a feeling of «pins and needles» in the hands and feet (neuropathy)".

Indeed, Asif is not alone in highlighting the health effects of these chemicals. Siddiqui, W. M. and Dhua, R. S. (2010) in 'Eating Artificially Ripened Fruits is Harmful' also postulate that, "... indiscriminate use ...on different types of fruits can lead to toxic effects. Due to lack of awareness and education people consume chemically ripened fruits". And again, another renowned scholar, Dhembare, A. J. (2013), in his paper, Bitter Truth about Fruit with Reference to Artificial Ripener also joins the chorus in highlighting the harmful impacts of these chemicals and states that, "If pregnant women consume fruit ripened with carbide, the children born could develop abnormalities".

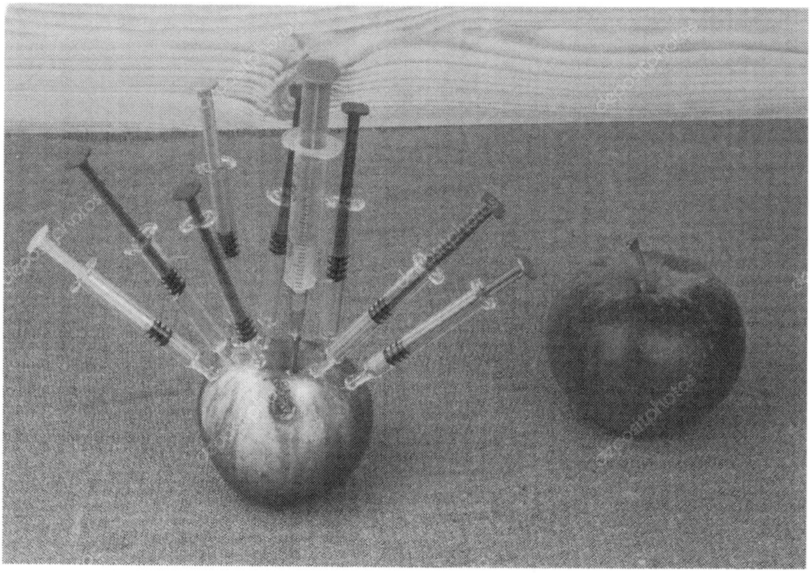

Figure 8.2: Fruit ripen by chemicals usage.

http://st2.depositphotos.com/2082539/8549/i/950/depositphotos_85494540-stock-photo-genetic-modification-concept-fruit-and.jpg

Dhembare however goes beyond Carbide and talks about the negative impacts of other chemicals used in vegetable growth, and fruit ripening and states that, "Ethylene is harmful which cause damage to the neurological system, affects the eyes, skin, lungs, memory and leads to prolonged hypoxia. It is a plant growth regulator". Again, he argues that, "There are some proves that oxytocin promotes ‹tribal› behavior, incorporating the trust and empathy of in-groups with their suspicion and rejection of outsiders, in individuals with autism spectrum disorders". Lastly, he states that, "Potassium dihydrogensulfate poses health hazards such as diarrhea, nausea, vomiting, cramps, and slow heartbeat; accelerate breathing, muscle weakness, paralysis".

Thus, one can deduce from this that, in as much as the need exists to find processes to feed the world large population, the chemicals used as to make vegetables large, pesticides and fruit reopeners may not always be safe. Perhaps, this explains why there was hue and cry when the Plant Breeders bill was laid in the Ghanaian parliament in 2013. When news broke about the Bill which had been in the Ghanaian parliament since

2013 in the latter part of 2013 and early 2014, the media latched on it and several varied views were expressed. One of such views is the one put forward by Food Sovereignty Ghana which has been captured on their website, foodsovereigntyghana.org. They state that, "Genetically Engineered (GE) crops, genetically modified organisms (GMO's) are questionable, their nutritional concentration is questionable, future and generational effects are unknown and the toxic chemicals that 99% of commercial GE crops poison us all".

Some of these concerns prompted no mean a person like the Speaker of Ghana's Parliament to wade into the controversy and on Tuesday, 28th January, 2014, the Daily graphic news portal reported on their website that, the speaker of parliament in his opening speech stated that, "he had received 'a number of petitions' on Ghana's Genetically Modified Organisms (GMO) agenda", and thus reminded the house that, "the Bio-safety Act, 831, which was enacted in 2011 made provisions for the establishment of a regulatory body that is the Bio-safety Authority to deal with most Bio-safety Authority to deal with most of the concerns raised in those petitions". Some view on the matter is that, in as much as the need exists to feed the worlds ever increasing population and the fact that fruits and vegetables generally have short life spans, which needs action if we are to preserve food, human health should be paramount in any decisions that allow or disallow the use of these chemicals in food production. Whatever benefits these chemicals provide, if their side effects are established to be dangerous, then they should not be encouraged and banned.

Thus more action needs to be taken beyond just banning the use of these chemicals, but also punishing those who use them and instituting agencies or bodies that will regulate its use in places where they are not banned. Yemen Times reporter Dawood, A. for instance reports that, in Yemen, "The Ministry of Agriculture has laid out guidelines for a safe harvest period, 10 days after the crop has been sprayed with pesticides, but farmers regularly ignore this regulation due to lose enforcement. Also, due to low oversight, consumers are often left in the dark about potential health risks related with chemicals that pose rapid ripening".

IN JUNK FOOD HAZARDOUS CHEMICALS OTHER THAN MSG

In fast food, MSG or Monosodium glutamate is far as of the simply chemical preservative consumed. Restrictive ingestion of junk food decreases the contact to many additional dangerous materials, like artificial dyes as well as trans fats. Moreover it reduces the general salt, calories as well as sugar intake.

Terrible Trans Fats

These chemicals are still there in several fast-food desserts as well as fried stuff, though several fast-food eating places are reducing their utilizaion of trans fats. By accumulating hydrogen to unsaturated fats, trans fats are prepared that means they contain an increased shelf life than additional sort of fat. These enhance lesser good cholesterol as well as bad cholesterol concentrations, a mixture which is dangerous to heart fitness. These also enhance swelling that is related to diabetes, stroke, as well as cancer and heart ailment.

Figure 8.3: Fast foods having all the dangereous chemicals.

http://i1142.photobucket.com/albums/n610/HCAppState/fastfoods.jpg

Bad Butylated Hydroxyanisole

BHA or Butylated hydroxyanisole is a chemical becomes the part of food which has oils to help stop them as of getting rotten. However, the chemical is a carcinogen that means it has the possiblity to cause malignancy. Animal learnings propose a well-built association among the chemical as well as the development tumors of cancer as well as also these human studies recommend so as to, it enhances the chance of stomach melanoma. BHA is present in provisions fried in living things fats, dehydrated potatoes, baked goods, and chicken, sausage as well as meat foodstuffs.

Horrifying Hydrolyzed Vegetable Proteins

Hydrolyzed vegetable proteins have a huge amount of salt, having monosodium glutamate as well as they are substandard sources of protein. The chemical is consumed to provide a fleshy taste to definite junk foods, as well as it also assists fast-food eating places decrease on how greatly of the definite foodstuff they utilize in their techniques. To estimate the long-term sound effects of the chemicals, further learnings ought to be made, especially in children.

Artificial Flavors and Dangerous Dyes

Brightly colored desserts as well as soda that are some junk foods have artificial flavors as well as dyes. Fruits, eggs and vegetables are mostly utilized to substitute actual provisions. Some colorings like Yellow #5 as well as #6, are possible carcinogens. Dyes may moreover enhance the hazard of performance disruptions as well as hyperactivity within offsprings. Sweeteners as well as artificial flavors as well include the possiblity of carcinogens.

The Normal Junk Food

Like to use each and every one of the poisonous chemical ingest fries a burger in addition to a cold drink. The burger probably have hydrolyzed vegetable proteins as well as BHA, along with numerous fast-food eating places still make their fries in oil so as to have trans fats. A colored beverage, have synthetic coloring, in addition to if it's fruit-flavored beverage, there is also a dose of synthetic essence. Substitute the burger in favor of chicken nuggets, furthermore still have hydrolyzed vegetable proteins as well as BHA. A pizza, in the crust there are trans fats, as well as Chinese food contains BHA, MSG as well as, depending on the investiture, flavors as well as artificial dyes.

PROCESSED MEAT

FOOD INGREDIENTS IN PROCESSED FOOD

Any food that has been canned, dehydrated, or had chemicals added to it is a processed food, and these foods make up about 60 percent of the average American diet.

By "toxic," mean chemicals or highly processed ingredients that aren't good for you or can pose harm to your health. Understand which toxic food ingredients to avoid:

1. Palm Oil

When a regular fat like corn, soybean, or palm oil is blasted with hydrogen and turned into a solid, it becomes a trans fat. These evil anti-nutrients help packaged foods stay "fresh," meaning that the food can sit on the supermarket shelf for years without ever getting stale or rotting. Eating junk food with trans fats raises your "bad" LDL cholesterol and triglycerides and lowers your "good" HDL.

2. Shortening

Ditch any food that lists shortening or partially hydrogenated oil as an ingredient, since these are also evil trans fats. In addition to clogging your arteries and causing obesity, they also enhance risk of metabolic syndrome.

Figure 8.4: Processed meat and food additives in fast foods.

https://images.vat19.com/covers/large/gummy-fast-food-2.jpg

3. White Flour, Rice, Pasta, and Bread

When a whole grain is refined, most of its nutrients are sucked out in an effort to extend its shelf life. Both the bran and germ are removed, and therefore all the fiber, vitamins, and minerals.

4. High Fructose Corn Syrup

The evil king of all refined grains is high fructose corn syrup (HFCS). The amount of refined sugar consume has declined over the past 40 years, but consuming almost 20 times as much HFCS.

5. Artificial Sweeteners

Aspartame (NutraSweet, Equal), saccharin (Sweet'N Low, SugarTwin), and sucralose (Splenda) may be even harder on our metabolic systems than plain old sugar. These supposedly diet-friendly sweeteners may actually be doing more harm than good.

6. Sodium Benzoate and Potassium Benzoate

These preservatives are sometimes added to soda to avoid mold from growing, but benzene is a known carcinogen that is also associated with serious thyroid damage. Dangerous concentrations of benzene can build up when plastic bottles of soda are exposed to heat or when the preservatives are combined with ascorbic acid (vitamin C). Don't risk it, people.

7. Butylated Hydroxyanisole (BHA)

BHA is another potentially cancer-causing preservative, but it has been deemed safe by the FDA. Its job is to help avoid spoilage and food poisoning, but it's a major endocrine disruptor and can seriously mess with hormones. BHA is in hundreds of foods. It's also present in food packaging and cosmetics.

8. Sodium Nitrates and Sodium Nitrites

These two different preservatives are present in processed meats like bacon, lunch meat, and hot dogs. They're some of the worst offenders, and they're believed to pose colon cancer and metabolic syndrome, which can lead to diabetes. Health must be always protected by choosing fresh, organic meats.

9. Blue, Green, Red, and Yellow

The artificial colors blue 1 and 2, green 3, red 3, and yellow 6 have been linked to thyroid, adrenal, bladder, kidney, and brain cancers.

10. MSG

Monosodium glutamate is a processed "flavor enhancer." While glutamates are found in some natural foods, such as meat and cheese, the ones exploited by the processed-foods industry are separated from their host proteins through hydrolysis.

CHOCOLATES

Chocolate is a food that contains a range of nutrients--including not only fats and sugar, but also other carbohydrates and proteins. In addition, chocolate consists of small quantities of salts of metals such as magnesium, potassium, calcium, and iron; the vitamin riboflavin; the stimulant caffeine; and water

The Sweet Lure of Chocolate

Figure 8.5: Chocolate.

http://www.saruralwomensgathering.org.au/wp-content/uploads/2017/07/chocolate.jpg

Then one must know the chemistry behind the chocolate making and why chocolate is considered as the feel-good factor, mood-alternating factor etc. Chocolate consists of more than 300 - 500 known chemicals, some of which react within the human brain to alter mood.

Out of these 300 - 500 chemicals the following chemicals play main role in humans:

Theobromine - Natural cough medicine. In large doses it may pose nausea and anorexia and that daily intake of 50-100 g cocoa (0.8-1.5g theobromine) by humans has been associated with sweating, trembling and severe headache. The theobromine LD50 is about 1000 mg/kg

Caffeine - chocolate consist of 27 mg of caffeine.

Phenylethylamine - it gives the pleasurable sensations and it is released by the brain when people are falling in love.

One of the reasons chocolate is unique is the temperature at which it melts between 94 °F and 97 °F by the same time the human body, at 98.6 °F is just above the chocolates melting temperature chocolate slides across the tongue and liquefies into a perfect puddle of taste feeling. Chocolate is a native food of South America. Beans of the Theobroma cacao tree are dried, shelled, fermented and ground with sugar, fats and other flavorings to form the wide variety of dark and milk chocolate.

Chocolate›s varied flavors, colors, shapes, and textures result from different recipe traditions that have evolved in various parts of the world. The essential ingredient in all chocolate is cocoa.

Chemicals in Chocolates

Types of chemicals found in chocolate.

- Xanthines
- Caffeine
- Theobromine

Phenolics - acts anti-oxidant & fat reducer.

Phenylethylamine - it is a 'love drug' and it gives pleasurable effects by releasing b-endorphin.

Anandamide - produces transient sensation of well being.

Serotonin - feel good factor in chocolate.

Fats and Sugars - these controls blood cholesterol and produces energy. The high-fat amount of most chocolate - Cadbury's Dairy Milk alone contains 30 g of fat per 100g these amounts of fat may pose obesity which intern leading to heart disease and diabetes.

Other Chemicals

Histamine - It is responsible for the craving of chocolate and change in mood.

Theophylline - Central nervous system and cardiovascular stimulant.

Chocolate Flavoring

Table 8.1: Group of chemicals gives flavor to the chocolate

No.of chemicals	Chemical functional group
47	Hydrocarbons
28	Alcohols
24	Aldehydes
41	Ketones
57	Esters
21	Nitrogenous compounds
15	Sulfur compounds
53	Acids
7	Phenols
14	Pyrrols
9	Pyridines, crinoline
80	Pyrazines, quinoxalines
7	Thiazoles
11	Oxazoles
34	Furans, pyrones, lactones

Health Benefits from Chocolate

Recent studies have shown that cocoa or dark chocolate has potential health benefits for people.

Dark chocolate is full of the flavonoids, epicatechin and gallic acid, which are antioxidants that help

- Protect blood vessels.
- Cardiac health.
- Prevent cancer.

It also has been effectively demonstrated to counteract mild hypertension.

ICE CREAM OR ICE CHEMICALS

Unfortunately, those days are gone when ice creams were made with beneficial products. Some companies are doing everything they can to decrease their costs of production and increase profits.

Ice cream is a huge favorite for lots of people, especially children.

Figure 8.6: Ice creams or ice chemicals.

https://static1.squarespace.com/static/52536cbde4b04bbc53b210f3/t/526a1942e4b0ca2844ff3aae/1382685033780/ice-cream+2.jpg?format=500w

Depending on the country, manufacturer, and product, analysis has shown that some if not most of the following chemicals are in most of the commercial ice creams.

Diethyl glycol (DEG) - Is used as a paint solvent and to make polyester resins and plasticizers. It's also used in anti-freeze for engines and paint removers. It's used as a cheap substitute (emulsifier) instead of eggs in ice cream to thicken the product.

Piperonal - also known as heliotropin, is used in place of vanilla as a cheap substitute. It's described in the National Library of Medicine

HSDB Database as "moderately toxic" as well as a "human skin irritant". It's used in head lice formulas to kill lice. No matter how well you sugar coat it - "moderately toxic" is still toxic!

Ethyl Acetate (EtOAc or EA) - Is used as a pineapple flavor. It is also used as a cleaner for leather and textiles. It's used in paints as an activator or hardener as well as being used in some glues and nail polish removers. Its vapors have been known to pose chronic lung, liver and heart damage.

Aldehyde C-17 - Is used as a cherry flavor. It is an inflammable liquid also used in aniline dyes and the manufacture of plastic and rubber.

Amyl Acetate - pentyl acetate, also known as banana oil is used as a flavoring agent mainly as banana flavor. It's also used as a paint and lacquer solvent.

Butyraldehyde (butanal) - Derived from butane, is used in the production of plasticizers, alcohols, solvents, polymers, rubber accelerators and cement, textile auxiliaries, perfumery, and flavors.

Benzyl Acetate (phenylmethyl acetate) - Used as a fragrance and food flavoring agent to name but one of its uses. It's also used in detergents, soaps, inks, polishes, lacquers and as a solvent in plastics and resins.

Propylene Glycol (1,2-propane diol or propane-1,2-diol) - Is used in food colors, as humectants (E1520) and preservative in food and for tobacco products. It's also a solvent in plastics and paint, used in aircraft deicing fluid, automotive engine antifreeze, liquid household and dishwashing detergents, as well as industrial soaps and cleaning fluids. The critical oral toxicity of propylene glycol is classed as very low. However, low or not, it still is toxic.

Sodium benzoate (E211) - Is mainly used as a preservative in foods, drinks, medicines and cosmetics. Just a preservative, it sounds so innocent. It's also used in fireworks as a fuel. It's been shown in studies to be a carcinogenic substance. Professor Peter Piper, a professor of molecular biology and biotechnology at Sheffield University, in 2007 said "These chemicals have the ability to pose severe damage to DNA in the mitochondria to the point that they totally inactivate it; they knock it out altogether. The mitochondria consumes the oxygen to give you energy and if you damage it – as happens in a number of diseased states – then the cell begins to malfunction very seriously. And there is a whole array of diseases that are now being tied to damage to this DNA – Parkinson's and quite a lot of neuro-degenerative diseases, but above all the whole process of ageing."

Also when it's combined with vitamin C or E, this combination causes benzene to occur. Benzene has been shown to pose cancer.

Potassium sorbate (E202) - is used as a preservative to prohibit the growth of yeast and molds to increase shelf life. It's been shown to be a respiratory, skin and eye irritant. Some studies have shown that it's both genotoxic and mutagenic to human blood cells. Potassium sorbate is found to be toxic to human DNA and that it negatively influences immunity.

Polysorbate 80 - E433, as well as many other names, are used as an emulsifier in foods, specifically in ice cream. Also used in soaps and shampoos. It's been shown to suppress the immune system and it can pose severe allergic reactions including anaphylactic shock.

Carboxymethyl cellulose, Monoglycerides, diglycerides and triglycerides, Disodium Phosphate, Mono stearates, Modified corn starch, Soy Lecithin, Calcium sulfate, Magnesium hydroxide etc.

MEAT

Every time you sit down to a meal containing meat, eggs or dairy products, you're dining on known carcinogens, bacteria, and other contaminants that can cumulate in your body and remain there for years. Animal products are loaded with bacteria, antibiotics, hormones, dioxins, and a host of other toxins that can pose serious health problems in humans. An extremely high percentage of all the flesh from the chickens, turkeys, cows, fish, and pigs butchered every year in the U.S. is contaminated with E. coli, campylobacter, listeria, or other dangerous bacteria that live in the intestinal tracts, flesh, and feces of animals.

But that's not the only danger related with dosing animals raised for their flesh with antibiotics. Roxarsone, an antibiotic usually used on factory farms, consists of significant amounts of the most carcinogenic form of arsenic. USDA researchers have found that "eating 2 ounces of chicken per day—the equivalent of a third of a half of a boneless breast—exposes a consumer to 3 to 5 micrograms of inorganic arsenic, the element's most toxic form." Daily exposure to low doses of arsenic can dramatically enhance the risk of cancer, dementia, neurological problems, and other ailments in humans.

Dangerous hormones are also lurking in meat and milk. According to the USDA, by 1999, roughly 99 percent of cows on large feedlots in the U.S. were given synthetic hormonal implants. These hormones, which make cows grow larger and produce more milk than they would commonly, are prohibited for over-the-counter use by humans in the United States. Scientists have clearly shown that these hormones can enhance the risk of disrupted development and cancer in humans.

A powerful hormone-disrupting group of chemicals called "dioxins" binds to a cell and modifies its functioning, potentially posing a wide range of effects, including cancer, depressed immune response, nervous system disruptions, miscarriages, and birth deformities. Researchers at the EPA have found that people who consume even small concentrations of dioxins from meat and dairy products have an extra one in 100 risk of suffering from cancer—solely as a result of their dioxin consumption and on top of all other risks.

Pesticides may also found in large quantities in meat and have been associated with a wide range of health problems in humans, including birth defects and cancer. Although humans do ingest some herbicides and pesticides from plant foods, scientists report that animal products are responsible for roughly 80 to 90 percent of dietary pesticide exposure. Pesticides are sprayed on crops that are eventually fed to farmed animals, and like dioxins, pesticides accumulate in their bodies over time. When we eat the flesh, milk, or eggs of animals, the pesticides that they have eaten during their lives are transferred to us.

Some chemicals that are used in today foods and can pose serious illnesses in living and nonliving.

CHAPTER

9

ESTIMATION VALUES AND LEGISLATIONS FOR CHEMICALS IN FOOD SUBSTANCES

SCOPE OF THE GUIDELINE FOR FOOD IN CONTACT WITH CHEMICALS (MOSTLY METALS)

The elements of periodic classification might roughly be separated into three categories: metals, metalloids as well as non-metals. Approximately 80% of all the essentials are consists of metals. Metals may typically be characterized on the foundation of their physical as well as chemical properties in the solid utter. A pure metal usually:

- Be a fine conductor of heat and electricity
- Due to superior reflectivity shows metallic gleam
- It may be stroked into sheets otherwise pressed permanently out of nature exclusive of breaking or cracking i.e. is malleable
- It can be altered into a skeletal cable i.e. is ductile
- Able to be merged or else melted.

Further characteristics of metals are so as to they readily form cations, form oxides, hydroxides and binary metal hydrides, because of the presence of one to five electrons in the remotest shield and be able to retort with different non-metals frequently giving ionic compounds in addition to among further metals forming metallic compounds. These metallic compounds may be referred to as alloys, comprised of numerous metals.

Metalloids fall in among two categories, as they have equally metallic as well as non-metallic distinctiveness. Instances of these are antimony and arsenic.

This rule covers the metalloids as well as metals so as to be exclusively used in FCMs otherwise present like impurities in e.g. alloys used in FCM. The elements incorporated are copper, aluminum, lead, antimony, chromium, arsenic, thallium, barium, magnesium, cadmium, lithium, mercury, titanium, zinc, cobalt, nickel, beryllium, tin, manganese, vanadium, iron, silver and molybdenum.

FOOD EXPOSED METALS AND RESOURCES

Metals along with alloys typically for route tackle in the food trade in addition to in personal households are used for a lot of kinds of provisions contact supplies. Boilers, pipes, spray dryers; pans, kitchen knives, pots, and storage containers crockery are several instances of method utensils as well as built-up foodstuff industrialized tools. Foils, cans Furthermore coating in multilayer substances, e.g. a metal layer in the plastic casing are examples of other uses of alloys and metals as FCM.

LEGISLATION FOR FOOD EXPOSED EQUIPMENT

FCM, regulation 1935/2004, the balanced guideline on victuals exposed equipment, coat ordinary demands for the entire forms of FCM.

To FCM the regulations are valid:

"Are deliberated to be brought into contact with food, are already in contact with food and were intended for that purpose, can reasonably be expected to be brought into contact with food or to transfer their components to food under normal or foreseeable situations of use."

GENERAL LEGISLATION AND EU DEFINITE DEALINGS

EU has exacting procedures for plastic (virgin as well as used), dynamic Moreover bright covering, earthenware, epoxy derivatives also revived cellulose, in addition, the common necessities for all categories of FCM.

The EU guidelines on FCM simply have need of a DoC for equipment among a precise appraise, e.g. plastic. As well the systemized guideline, a number of countries, takes account of Denmark, Norway as well as several Nordic countries. Denmark and Norway have loaded for an affirmation of acquiescence, DoC, for the entire nature of resources, together with metals, cardboard, and paper. Moreover, Nordic countries have state legislation on the liberation of metals from FCM, e.g. from tumbler and earthenware.

In regulation 882/2004 on authorized controls, regulations in favor of provisions establishments, the demands are integrated whereas in regulation 2023/2006 for companies and FCM, regulations are expressed. GMP, good producing practice characterizes on Regulation 2023/2006 "means those aspects of quality assurance which ensure that materials and articles are incessantly manufactured and controlled to ensure conformity with the regulations applicable to them and with the quality standards suitable for their intended use by not endangering human health or posing an unacceptable variation in the composition of the food or causing a deterioration in the organoleptic characteristics thereof." Furthermore, it declares with the intention of feature assertion "means the total sum of the organized and documented arrangements made with the purpose of ensuring that materials and articles are of the quality demanded to ensure conformity with the rules applicable to them and the quality standards necessary for their conscious use."

CERTIFICATION OF CONFORMITY, DOC (DOCUMENTATION OF COMPLIANCE)

Citations comprise of the sustaining credentials in addition to the declaration of conformity, DoC. The domestic organize in production, as well as deal, must be acknowledged as segments of rules with conformity.

The assertion of acquiescence is used amongst deal associates to affirm with the purpose of the provisions exposed stuff conforms with the system appropriate to it. These ways so as to the answerable business have to manuscript observance created it accessible to their clients plus with the applicable policies. Its principle is to fortify the organization in addition to liability of the dealers at all steps of assembly, plus that of the preliminary essences.

The sustaining documents can be reserved secretly, but ought to be presented for the establishment on demand, along with it's frequently use in production plus deal, e.g. when mentor are inspecting for acquiescence. The underneath certification is the absolute records of fulfillment, which composed of every single one information, e.g. recognition sum of means and on logical taxing practices.

Second-hand construction plans as these may be used for decades as well as may have components substitute through this duration may present confront regarding documents of conformity. Applicable records should normally be accumulated for 5 years as well as stretched as the apparatus is in the exploit, intended for such tackle. Though, when parameter 1935/2004 went into influence, conformity should not be renowned for used construction paraphernalia purchase aforementioned to 3rd December 2004. Barely, including adding up novel components such as e.g. pipes, hoses and gaskets must be documented, observance of maintenance prepared later than this meeting.

In 2015, this work is predictable to be concluded. Nordic exertion on a test catalog for sketching the statement of fulfillment for the ultimate provisions exposed material is constant.

VERIFICATION

Traceability concerned to every single one elements of FCM, including additional fractions for assembly utensils. All corporations alongside the bring in the sequence are responsible for manuscript identification and verification, i.e. counting victuals maker, importers, distributors as well as traders of FCM preparatory stuff, intermediates in addition to ultimate FCM. These supplies went into potency in October 2006.

Traceability ought to be accessible commencing the trader to the consumer – "one step back and one step forward." For illustration, a plastic producer must comprise information regarding the synthetic tablet dealer in the credentials for the finishing FCM; in addition, a sandwich ingot must have certifications and traceability for causing moreover containers utilized for exposed food.

TECHNOLOGICAL CHARACTERISTICS

Technological characteristics of FCM, like e.g. construction machinery, are the consistency of the provisions manufacturers. The law on sanitation of chow, guideline 852/2004 locates out hassle in favor of FCM. Agreement with the lawful necessities on fortification and organoleptic inertness in addition to, while deciding on progression, utensils and further varieties of FCM tranquil of alloys Furthermore metals, concern resembling appropriateness for the suitable utilization, foodstuff exposed, pH, heat, upholding necessities with sterile distinctiveness, should be taken interested in relation. Three victuals manufacturers are liable for sanitary storage as well as managing of FCM in regulation to obstruct traverse contagion. For development apparatus, particularly the application of sterilizer can be a dispute for explicit alloys.

Rusting of the plane will diminish the potential of a fine disinfected category of the FCM, for metals as well as alloys.

VICTUALS CONTAMINATION BY METALS

The law on contaminants in provisions, Regulation 1881/2006, takes description of the consequent metals: lead, inorganic tin, cadmium as well as mercury by every single one basis, counting liberation from FCM. Confines for the metals are deposit in numerous varieties of foodstuff as well as, necessities for analysis; samplings along with exposure, in addition, to be taken into description when appraising latent discharge from FCM, are involved in this legislation in regulation 333/2007.

This can be complex in the construction of FCM as the ecological surroundings corruption of the food factors might not be recognized. Still, under observation, e.g. lead is frequently instituted in the cast-off glass. Though, in common, cadmium, mercury in addition to lead should not be there in FCM.

Regulation Standards for Metal Ions Discharge

The summary of regulation standards intended for liberation of metal ions to provisions are deliberated to assist official victual supervisor, production along with deal in serving conformity of FCM. Nevertheless, it is tinted so as to in the authorized organize; the sum should be described like supervision merely, as well as definite FCM more than one of the assistance amounts should be estimated on a case-by case supply.

HAZARD ESTIMATION

In EU system 1881/2006 on contaminants Locating control principles utmost confines in favor of inorganic tin, mercury, lead, and cadmium in several natures of provisions are a deposit. Consequently, these standards are highest confines irrespective of the cause of the contagion as well as contain discharged from FCM. Like an exemplary, highest values for inorganic tin are put for preserved sustenance, where the source discharges from containers. The confines intended for tin in preserved chow modify as of 50 to 200 mg/kg.

Usually, control standards are standing on global hazard measurement from the European Food Safety Authority, EFSA, as well as/otherwise the WHO/FAO Joint Expert Committee on Food Additives as well as Contaminants, JECFA. The EFSA menace measurement information is available in the course of the associated particular in factor on the entity metals. As support for the regulation standards, these hazard review information may be utilized. By means of the **"ALARA principle,"** the systematic threat reviews by EFSA terminate in various points that a compound should be managed in agreement. This means that the substance in provisions ought to be "As Low As Reasonably Achievable." Highest values for the substance in provisions might thus be supported on what is logically attainable. In favor of case in point, the concentrations designed for lead is in the series from 0.02 to 1.5 mg/kg.

For the construction and exploit of FCM, the GMP necessities must formulate so as to the ALARA principle is reserved. In the detailed EU-regulations on FCMs prepared of artificial, commandment 10/2011, precise most relocation confines in favor of essence are stand on the tolerable daily intake, TDI, of the material as well as the possibility to facilitate a 60 kg individual devours 1 kg of enclosed provisions for every day. This is the point for various phthalates for e.g. DEHP, where just 50% of the TDI is owed for scheme the definite upper limit for passage from FCMs prepared of plastics. During numerous cases, the confines take into description introduction since further start e.g. ecological contagion. The Council of Europe utilizes similar contemplations intended for locale specific release limits, SRLs, in their supervision manuscript on metals as well as alloys.

UNIVERSAL SUPPLIES

By means of certain EU laws, FCM consists of metals as well as alloys may not enclose. Thus "The materials and articles shall be manufactured in compliance with good manufacturing practice so that, under normal or foreseeable conditions of use, they do not transfer their components to food in quantities which could:

- endanger human health
- bring about an unacceptable variation in the composition of the food
- bring about the deterioration in the organoleptic characteristics thereof."

FCM DISCHARGE METAL IONS CONTROL STANDARDS

Discharge of metal ions by alloys ought to be reviewed separately in favor of the entire elements utilized in the alloy as the liberation of the isolated elements. The subsequent concerns ought to be prepared while setting confines used for discharge of metal ions from materials exposed to food:

MAIN COMPONENTS AND IMPURITIES

Aluminum (Al) – Main Component

What and Where

Aluminium is found in many minerals as well as the third plentiful constituent in the earth's crust. Besides FCM, aluminum may be capable of the component to the absolute individual introduction because of utilization in various sustenance additives, particularly in Ensign Moreover in antiperspirants.

0.2 to 1.5 mg/kg bw/week is usual nutritional intakes assortment for aluminium.

Toxicological Effects

Just a slight quantity of aluminum is captivated other is excreted through the kidneys, as well. Various aluminum users confirm compassion towards aluminum. By EFSA, the supportable on a daily basis ingestion place is found having sound effects on the emergent head, neuro in addition to rudiment toxicity.

Risk Assessments

In 2008 EFSA has put a PTWI that is the provisional tolerable weekly intake of 1 mg/kg bw/week. The PTWI may enhance the ingestion.

In 2011 JECFA has put a PTWI of 2 mg/kg bw/week.

Europe Council Limit Value

In 2013, on the basis of ALARA principle, the Europe Council has placed a definite liberation value 5 mg/kg.

Other Limit Values

Decree (98/83) stipulates a quantity of 200 µg/L in the drinking water.

Nordic Comments

Though, the normal concentrations found into victuals can be elevated contrast to discharge as of victuals exposed equipment as well as obscure a lesser liberate value. 10% of the EFSA PTWI keeps up a correspondence to 1 mg/kg.

Analysis

Locale contagion all through examination is extremely general. In provisions, the edge of quantification through ICP-MS is around 0.1 mg/kg.

FCM

As it is frequently a constituent in alloys. Aluminum is intended for e.g. sauce pans, coffee container Furthermore preserve, as well.

Developed Food

Industrially processed provisions contain aluminum by way of victuals enhancers.

Recommendations

Storage of aluminum should be avoided in alkaline, acidic otherwise salty sustenance, particularly fluids, is exposed. For users, uncoated aluminum must be tagged. An illustration of such a sticker might be there:

"User information: do not use this utensil for storage and processing of acidic, alkaline or salty food."

Antimony (Sb) – Main Component

What and Where

In addition to individual action, antimony is discharged keen on the atmosphere as of normal.

From 0.03 to 0.06 μg/kg bw/day is the characteristic dietetic ingestion series.

Toxicological Effects

Antimony poses decreased stiff mass get, victuals along with fluid ingestion into the investigational natural world.

Risk Assessments

In 2004 EFSA has predictable antimony trioxide along with support on the WHO evaluation planned a restraint of 0.04 mg/kg of victuals as antimony. In 2008 and 2011 WHO has put a TDI of around 0.006 mg/kg bw/day.

Europe Council Limit Value

In 2013, the Europe council has put a definite liberate perimeter of 0.04 mg/kg allotting 10% of the toxic suggestion assessment put via WHO of 0.36 mg/day as well as suppose a 60 kg person overwhelming 1 kg per day of provisions.

Other Limit Values

Directive (98/83) indicates a maximum value of 5 μg/L, the EU ingestion fluid. European set EN 610/1995 for tin as well as tin alloys has put a definite movement maximum value in favor of antimony of 0.01 mg/kg. In 2004, the European Commission has put an exact movement frontier, SML, of 0.04 mg/kg in favor of antimony inside synthetic provisions contact equipment (rule 10/2011).

Remarks of Nordics

No comments.

Analysis

In food, the maximum value of quantification through ICP-MS is almost 0.001 mg/kg.

FCM

The manufacture of lead, like a fire-proofing apparatus in the fabric as well as plastic, tin as well as copper alloys, like a dye in the coating, in addition to like a pacifying mediator in the glass, earthenware moreover enamels are formed by antimony.

Raw Food

The majority food excluding milk, eggs, oil, and fats has antimony in it. Maximum quantities are found in sugar, meat as well as fish.

Processed Food

In cakes as well as chocolates have highest amounts.

Arsenic (As) – Impurity

What and Where

Arsenic may be present as organic otherwise inorganic arsenic. In drinking water and rice, arsenic is set up like inorganic arsenic, while in seafood and fish arsenic is originate like organic arsenic. Inorganic arsenic has adverse sound effects. 0.13 to 1.22 microgram/kg bw/day has specific food ingestion of inorganic arsenic assortment.

Toxicological Effects

Severe venomous poses diarrhea, nausea with pain and blood in stomach and esophagus. Organic arsenic is not of fitness distress. Inorganic arsenic has carcinogenic effects in skin, lungs, kidneys plus bladder.

Risk Assessments

In 1989, JECFA had put a temporary tolerable weekly ingestion (PTWI) of 15 μg/kg bw/week. 0.3–0.8 μg/kg bw/day is standard dosage intensity established by EFSA. In 2009, EFSA anticipated inorganic arsenic as well as institute so as to there is health unease at intakes less than this as well as depart the PTWI.

Europe Council Limit Value

In 2013, the Europe council has put a definite liberate intensity of 0.002 mg/kg allocate 10% of the toxicological orientation value, as well as 0.0003 mg/kg bw/day assuming a person of 60 kg consumers.

Other Limit Values

The EU Drinking water specifies a frontier of 10 μg/L by order (98/83). There are no EU utmost limits for arsenic in food. Conversely, there are summary intended amounts for inorganic arsenic in rice and rice yield in EU and Codex.

Nordic Remarks

No comments.

Analysis

The significance of supposition of inorganic arsenic by HPLC-ICP-MS is almost 0.001 mg/kg in food, as well as around 0.01 mg/kg, is the maximum value of quantification for overall arsenic by ICP-MS.

FCM

In several metals as well as alloys arsenic may be present as an adulteration.

Raw Food

In drinking water and rice, inorganic arsenic may be present.

Processed Food

In rice biscuits/wafers and rice imbibe, inorganic arsenic is present.

Barium (Ba) – Impurity

What and Where

In definite metals and alloys, barium is at hand like a contamination. Especially dietetic ingestion assortment is from 0.4 to 2.7 mg/day an individual.

Toxicological Possessions

Toxicological Effects of barium are cardiovascular illnesses as well as enhanced hazards of nephropathy.

Risk Assessments

The Scientific Committee on foodstuff did in 1996 approximate barium in relation to normal mineral water as well as suggested a higher intensity of 1 mg/L in natural mineral water. US EPA has put a toxicological planned significance RfD of 0.2 mg/kg bw/day

Council of Europe Limit Value

In 2013 the Europe council has put a definite liberate maximum value of 1.2 mg/kg allow 10% of the US EPA toxicological suggestion importance RfD of 0.2 mg/kg bw/day as well as pretentious a 60 kg person overshadowing 1 kg of foodstuff.

Other Maximum Value

Regulation 10/2011 has the definite migration limit for synthetic food contact materials is 1 mg/kg. The directive 98/83 no specifies a limit in EU drinking water.

Nordic Interpretation

No remarks.

Analysis

In food, the value of quantification through ICP-MS is almost 0.09 mg/kg.

FCM

Color tumbler, earthenware, porcelain as well as an anti-corrosion pigment for metals barium chromate is utilized. In pigments, varnish in addition to glass barium chloride is used. As corrosion inhibitors in low-carbon steels, barium bromates are consumed. In addition, barium is used in the manufacturing of soap, paper, linoleum, and rubber.

Raw Food

Main origins in the human diet are milk, flour, potatoes and several nuts.

Beryllium (Be) – Impurity

What and Where

Beryls are the most abundant form of beryllium, in addition, is mostly established in the appearance of oxides or complex beryllium-aluminium-silicates. In Spain, the precise dietetic ingestion is assessed to be almost 20 μg/person/day.

Toxicological Effects

Inhalation toxicity is the most related toxic effect.

Risk Assessments

1998 the USEPA suggested a reference dosage, RFD, of 0.002 mg/kg bw/day Moreover a verbal tolerable ingestion of 0.002 mg/kg bw/day derived in the Concise International Chemical Assessment Documents, CICAD.

Europe Council Limit Value

In 2013 the Europe council has put a precise discharge concentration of 0.01 mg/kg assigning 10% of the passable ingestion of 0.002 mg/kg bw/day as well as insolent a 60 kg person overwhelming 1 kg of victuals.

Other Limit Values

No certain limit by Directive (98/83) in the EU drinking water.

Nordic Clarification

No remarks.

Analysis

In food roughly 0.007 mg/kg is the value of supposing through ICP-MS.

FCM

It is an outstanding heat player as well as extra squashy than steel. It is particularly used like a consolidating mediator in alloys as well it is opposed to rigorous nitric acid. As well, it is utilized for dental prosthesis as well as it can be found in alloys for shaping plastics in addition to investigate boilers and channels. In ceramics, beryllium oxide may be present.

Raw Food

Drinking water is the most important intake.

Cadmium (Cd) – Impurity

What and Where

In a variety of foodstuff, e.g. cocoa beans, sunflower seeds as well as in some mushrooms cadmium accrues which is a metal. Elevated amounts of cadmium may be present in Phosphate fertilizers. 2.3 to 5.4 μg/kg bw is a particular dietetic ingestion range per week. It is chiefly because of utilization of vegetables in addition to cereal foodstuffs.

Toxicological Effects

Kidney disruptions can be posed by the accumulation of cadmium inside kidneys.

Risk Assessments

2.5 μg/kg bw/week provisional tolerable weekly intakes (PTWI) have been described by EFSA. The PTWI is minute moreover accomplished that at hand is required to reduce disclosure to cadmium at the universal populace intensity in addition to EFSA articulated disquiet so as to the border among the usual weekly intake.

Europe Council Limit Value

In 2013 the Europe Council has put in order 98/83, a particular liberate maximum value of 0.005 mg/kg analogous to the frontier of drinking fluid. This maximum value matches to allotment of 25% of TDI of 0.00036 mg/kg bw/day as well as conceited a 60 kg individual overwhelming 1 kg of fare apiece day.

Further Frontier Standards

In regulation 1881/2006 EU has put utmost confines used for foodstuff. 0.005 mg/kg is anticipated to be the highest confine meant for the exodus. 0.01 mg/L otherwise 0.002 mg/dm2 is a liberation value for Norway. 0.1 mg/dm2 by FCM is a discharge value for Finland. Directive (98/83) stipulates a limit of 5 μg/L for the EU drinking water.

Annotations of Nordics

0.002 mg/kg of the PTWI would consequence in 10% of the EFSA.

Analysis

In food 0.001 mg/kg is almost the maximum value of calculation by means of ICP-MS.

FCM

Cadmium may be found in pewter.

Raw Food

In kidneys maximum concentrations may be found, sunflower seeds, cocoa beans as well as bivalve mollusks. In some vegetables as well as fruits, in developmental victuals cadmium cumulates, because of uptake via the roots from earth. As a result, amounts are not diminished through cleaning otherwise flaking. In zinc sulfate cadmium utilized as a provision more otherwise in favor of nourish, perhaps a dilemma.

Recommendation

For salty or else acidic sustenance, pewter must not be utilized. In cases wherever embellishments include a glazing having cadmium as well as may be falsely used for foodstuff exposed equipment, they must be marked example: "Releases cadmium should not be used for food or beverages".

Chromium (Cr) – Main Component

What and Where

In the environment, chromium takes place habitually in its trivalent appearance. As an industrialized impurity, the hexavalent structure is at hand. In 2014 EFSA estimates to normal nutritional ingestion variety as of 0.6 to 1.6 μg/kg bw per day designed for Cr(III).

Toxicological Effects

For toxicology, the speciation of chromium is of immense importance. Cr(III) is a major element on behalf of humans also is the main established status. Including effects like genotoxicity, Cr(VI) is extremely poisonous. Cr(III) is expected to present as discharge as of foodstuff exposed equipment.

Risk Assessments

For chromium (III) in 2014 EFSA has put a TDI of 300 μg/kg bw/day

Europe Council Limit Value

In 2013 the Europe Council has put a characteristic liberate concentration of 0.250 mg/kg supported on the WHO a 60 kg person overriding 1 kg of fare per day as well as utmost intensity for ingestion of 0.250 mg/person per day.

Additional Frontier Concentrations

A liberate quantity of 2 mg/dm2 in Finland is designed for FCM. Directive (98/83) stipulates a maximum value of 50 μg/L in the EU drinking fluid.

Interpretation by Nordics

No observations.

Analysis

About 0.03 mg/kg is the limit of observation by means of ICP-MS in provisions.

FCM

In containers, specific alloys, as well as tackles chromium, are at hand. As well, it is functional in metals commencing oxidization further like a varnish to shield, e.g. in canister chromium supplied to compliance, the tin coating.

Raw Food

Vegetables, cereals, white fish, as well as meat, are chief origins in the diet. Less than 0.1 mg/kg of chromium may be present in many natures of victuals.

Processed Food

Chromium(III) may be found in victuals enhancements.

Cobalt (Co) – Main Component

What and Where

It is an infrequent constituent. It describes for 0.05–0.1% of definite categories of the brace as well as is utilized for elevated potency alloys. 0.18 to 0.31 μg/kg bw per day is archetypal nutritional ingestions assortment.

Toxicological Sound Effects

In vitamin B12 cobalt is an essential constituent at hand. Cobalt toxicity poses effects on the abdominal pain, heart, blood pressure, as well as complexities while inhalation and in most horrible cases, fatality.

Risk Judgment

In 2001 the Dutch nationwide organization in support of communal fitness in addition to the surroundings, RIVM, has put a TDI of about 0.0014 mg/kg bw per day. In 2012 EFSA has expected cobalt complexes with respect to nourish.

Europe Council Limit Value

In 2013 the Europe council has set a certain release limit of 0.02 mg/kg allocating 20% of the toxicological reference dose, TDI at 0.0014 mg/kg bw/day, and assuming a 60 kg person consuming 1 kg of food per day.

In 2013 the Europe council has put a definite discharge concentration of 0.02 mg/kg allotting 20% of the poisonous suggested dosage, TDI of 0.0014 mg/kg bw per day, as well as presuming a 60 kg individual utilizing 1 kg of sustenance apiece daily.

Further Frontier Concentration

In 2003 by legislation 10/2011 the European commission has put the movement limit intended for salts of cobalt 0.05 mg/kg for synthetic materials exposed to food based on chance evaluation by RIVM.

No specific values specify by Directive (98/83) in the EU drinking water.

Nordic Interpretations

0.01 mg/kg is 10% of the RIVM, TDI.

Analysis

Almost 0.002 mg/kg is the limit of supposition through ICP-MS in food.

FCM

Manufacturing of steel, like a blue color in a tumbler, enamels as well as earthenware, in addition, to normalize the yellow color by iron in glass cobalt is utilized. In protective coatings on steel to enhance the adherence of the coating to the metal, cobalt oxide is used.

Raw Food

In food, cobalt is usually found in very low concentrations which are 0.01–0.5 mg/kg. Green leafy vegetables are a general source of cobalt.

Processed Food

In food supplements, cobalt can be found.

Copper (Cu) – Main Component

What and Where

In most types of sustenance, copper is usually present in the nature of copper ions otherwise salts of copper.

1.25 to 4.1 mg/person per day is specific dietetic ingestion series.

Toxicological Sound Effects

For humans, copper is an essential element. Poisonous causes are mainly critical causes in epigastric pain, nausea, vomiting as well as diarrhea.

Risk Assessments

In 2003 SCF as well as in 2006 EFSA have put an upper bearable limit for adults of 5 mg/person/day. For children age 1–3 years an upper value of 1 mg/day was set.

Europe Council Limit Value

In 2013 the Europe Council has specifies a typical libration quantity of 4 mg/kg allotting 39% of the harmful suggested dosage of 9.8 mg per day from the EU harm estimation value, RAR, as well as presuming a 60 kg individual utilizing 1 kg of victuals apiece day. Children were not supposed due to less contact with exposed provisions materials made by harmful metal that is copper.

Other Limit Values

By legislation 10/2011, the movement value for synthetic sustenance exposed substances is 5 mg/kg. Directive (98/83) specifies the quantity of 2 mg/L in the EU drinking water.

Nordic Interpretations

No remarks.

Analysis

In food substances, the concentration of calculations through ICP-MS is about 0.04 mg per kg.

FCM

In alloys, copper is utilized e.g. nickel silver, bronze in addition to brass. Copper utensils may be invaded by acidic food.

Substances composed of copper are naturally consumed into used for instance jam production, sweet, cheese making, chocolate as well as breweries. As well, few utensils, e.g. saucc-pans, may be composed of copper. Normally, these are coated with tin or stainless steel.

Raw Food

Pecans, offal, milk, meat, fish as well as green vegetables are the main source of copper by raw sustenance.

Processed Sustenance

Beer, cheese, jam, chocolate as well as sweets so as to be prepared in utensils composed of copper.

Recommendation

For sustenance, exposure doesn't use unprotected utensils of copper.

Iron (Fe) – Main Component

What and Where

In the earth's crust, iron is the fourth frequent element. In most kinds of sustenance, iron is present. 10 to 72 mg/person per day are specific dietetic ingestion series.

Toxicological Effects

Unsympathetic gastrointestinal sound effects (e.g. constipation, epigastric distress, nausea) have been described after instant verbal quantity at 50–60 mg each day of supplemental non-haem iron measures, particularly stipulation in use lacking victuals. The iron shortage is in general accredited quandary in addition to iron is a chief element used for the production of blood pigments.The unintentional toxin causes with therapeutic iron perceives serious harm of cardiovascular structures pancreatic, hepatic as well as gastrointestinal disruptions after intake of extremely elevated quantity. The severe dosage of 60 mg iron/kg bw may be fatal, other than oral amounts beneath about 10– 20 mg iron/kg bw do not create delicate universal toxicity.

Risk Assessments

In 1983 JECFA has put a temporary utmost acceptable day by day ingestion, PMTDI, by the side of 0.8 mg/kg bw/day. In 2004 EFSA has at hand so as to facts were inadequate to place a superior unobjectionable maximum value.

Europe Council Limit Value

In 2013 the Europe Council has situated a definite discharge assessment of 40 mg/kg, as this stage is logically possible.

Further Limit Values

By rule, 10/2011the fastidious immigration perimeter used for an artificial fare in touch resources is 48 mg/kg. As per directive (98/83) the EU consumption fluid stipulate an intensity of 200 μg/L like a scientific assessment.

Nordic Observations

4.8 mg/kg communicates by 10% of PMTDI.

Analysis

0.1 mg/kg in provisions is the frontier of calculation by means of ICP-MS.

GMP

Lids along with shutting for glass bottles in addition to jars have iron. Moreover, iron is the chief fraction of brace that is utilized for the e.g. canister, cast iron is consumed for vessel also pans, plus paint dyes also for the coloring of foodstuff iron oxides are utilized. Materials exposed to food prepared of iron might liberate elevated quantity of iron.

Raw Food

The highest quantities of iron are found in kidney, beef, liver, soya bean, egg yolk as well as ham.

Processed Food

In several states, wheat flour is prepared with iron to offer the suitable concentrations of iron within the food.

Lead (Pb) – Impurity

What and Where

From airborne contagion, dust or soil lead infectivity in foodstuff originates primarily. Especially, in fruit as well as vegetables with a large surface area to volume ratio, the highest concentrations are present like e.g. small barriers plus cale.

Previously lead has been used in paint, batteries, petrol, soldering cans for food also for water pipes. It is a metal which has typical dietetic ingestions assessments as of 0.36 to 2.43 μg/kg bw per day.

Toxicological Sound Effects

Particularly children are influenced as their rising nervous arrangement is susceptible to this metal. It disturbs the brainpower as well as the talent to be trained. Children take up extra lead from sustenance than adults.

Risk Assessments

In 2010 EFSA has reevaluated the chance of lead. EFSA found that lead was of additional toxicological disquiet than presumed in the past. The PTWI of 25 μg/kg bw/week was withdrawn, as well as EFSA suggested that no inferior quantity intended for endurable ingestion must be put. BMDLs distinct from blood lead intensity corresponding nutritional ingestion value were: developmental neurotoxicity BMDL01, 12 μg/L or 0.50 μg/kg bw/day; special effects on systolic blood stress BMDL01, 36 μg/L or 1.50 μg/kg bw per day; sound effects on occurrence of persistent kidney ailment BMDL10, 15 μg/L or 0.63 μg/kg bw per day. EFSA recommended that contact to lead, from equally dietary as well as non-dietary origins ought to be minimized. EFSA accomplished so as to it was impossible to eliminate the jeopardy to the rising fetus throughout introduction of pregnant feminine.

Europe Council Limit Value

In 2013 the Europe council has put a specific discharge frontier of 0.01 mg/kg equivalent to the limit place in drinking water as per order 98/83. Presumptuous so as to an individual of 60 kg using 1 kg of sustenance per day this keeps in touch to an allotment of 26% of the toxicological suggestion concentration as of BMDL10.

Further Limit Values

In rule 1881/2006 EU has utmost limits in provisions. The concentrations array is as of 0.02 to 1.5 mg/kg. Command (98/83) specifies a perimeter of 10 μg/L in the EU drinking water. 0.1 mg/L or 0.02 mg/dm2 is a release limit of Norway. The EU earthenware decree has currently being modified. Novel highest concentrations are possible to be 0.01 mg/kg. A release intensity of 0.5 mg/dm2 is for Finland.

Nordic Annotations

TDI would result in 0.004 mg/kg, 10% of the EFSA.

Analysis

0.01 mg/kg in food is the limit of observation by means of ICP-MS.

FCM

In pewter and tin, lead may be present as an impurity. Soldered procedure tackle, printing inks, and ceramics are a further source of lead. Furthermore, the old water pipes from the first half of the 1900s along with earlier than this can have the lead. Origination of lead in honey and spices has been traced from method apparatus. From various types of coffee machines lead release that has been revealed by recent German plus Swedish studies.

Raw Food

Water, vegetables, cereal products and fruit are the primary source of lead ingestion. Fruits in addition to vegetables with a huge plane region to volume ratio have superior concentrations. Considerable concentrations of lead are found in wild game shot with lead bullets.

Processed Food

Some teas and food supplements may have significant quantities of lead metal.

Recommendation

Storage of acidic provisions in sparkler glass will improve the discharge of lead so, don't store port wine otherwise supplementary acidic beverages in crystal glass while crystal glass have up to 24% lead. Glazing on earthenware may have cadmium or lead, which can transfer into the foodstuff thus, be careful with introduce of earthenware. Products labeled with the fork and glass representation should not pose any unfavorable liberation of lead. For food contact materials adornments may have its hazardous effects.

Lithium (Li) – Impurity

What and Where

It is not in metallic form because of its elevated reactivity. It is immensely dispersed on earth, although typically like a contamination in the salts of supplementary alkali metals. It is the lightest metal.

Characteristic dietetic ingestion is about 20–50 μg/person per day.

Toxicological Effects

For mood regulation, pharmaceuticals use salts of lithium.

Risk Assessments

In 1991, RIVM has placed a TDI of approximately 0.008 mg/kg bw per day.

Europe Council Limit Value

In 2013 the Europe Council has placed a specific liberate concentration of 0.048 mg/kg allotting 10% of the venomous mention quantity in addition to pretentious a 60 kg person overwhelming 1 kg of sustenance per day.

Further Limit Values

Guideline 10/2011describes the certain exodus concentration for artificial food contact equipment is 0.6 mg/kg.

The command (98/83) states no level in the EU drinking water.

Nordic Clarification

No remarks.

Analysis

Almost 0.006 mg/kg in food is the significance of observation via ICP-MS.

FCM

Lithium is utilized in low thermal-expansion glasses, ceramics as well as in alloys for high-quality mechanical components.

Raw Food

0.012–3.4 mg/kg is the concentration in food. Water is the main supplier to ingestion.

Magnesium (Mg) – Main Component

What and Where

In the earth's crust magnesium is the eighth mainly plentiful element, as well as salts of magnesium, are much copious in marine water. Approximately 300 mg/person per day are characteristic nutritional ingestion.

Toxicological Effects

For humans, magnesium is a vital constituent. Diarrhea poses when intake in great amounts happens. In cases of renal deficiency, poisoning can take place in children.

Risk Assessments

In 2001 SCF has put a better endurable quantity of 250 mg per day intended for eagerly dissociable salts of magnesium in dietary complement. In beverages and foods, this superior concentration doesn't comprise magnesium.

Europe Council Limit Value

In 2013 the Europe Council has determined not to place an exact discharge quantity as unfavourable special effects because of discharge from food exposed materials are doubtful.

Further Limit Values

Command (98/83) stipulate no concentration in the EU drinking fluid.

Nordic Explanation

Discharge of magnesium will generally be there greatly lesser than the concentrations present normally in foodstuff.

Analysis

Around 0.5 mg/kg in food is the concentration of observation through ICP-MS.

FCM

In alloys utilization of beverage containers, magnesium is typically utilized in aluminium-magnesium.

Raw Food

Followed by whole grains as well as cocoa, seafood has the uppermost quantities of magnesium.

Manganese (Mn) – Main Component

What and Where

In the majority of the foodstuff, manganese is present. Increasing future concentrations in food in some states, manganese has substituted organic lead like an enhancer in petrol. 2 to 7.4 mg/person per day range of nutritional ingestion.

Toxicological Sound Effects

In two universal oxidation natures manganese found, manganese (II) as well as manganese (IV). For humans, manganese is an important ingredient. By industrial revelation, sound effects on the vital nervous system as well as neurological impacts have been studied.

Risk Assessments

In 2006 EFSA accomplished so as to a greater frontier cannot be put as of restrictions into individual information. 0.28 mg/kg bw per day is observed to be the lowest severe effect level in growing male rats. In 2009 EFSA accomplished with the aim of the supplemental drinking and eating of 4 mg per day as well as 0.5 mg/day for aged citizens must not be surpassed.

In 2003 WHO has set a level of 0.06 mg/kg body weight per day, but in extracted this over again. The skilled assembly on vitamins as well as mineral deposits in the UK commission going on harmful effects of elements in victuals, customer foodstuffs Furthermore the surroundings, in 2003 COT EVM has placed a recommended intensity of 9 mg per day found on neurological special effects.

Europe Council Limit Value

In 2013 the Europe Council has placed a definite discharge quantity of 1.8 mg/kg consigning 20% of the poisonous verge of 9 mg/person per day place through COT EVM, as well as pretentious a 60 kg individual overwhelming 1 kg of sustenance daily.

Other Limit Values

Order 10/2011 the meticulous exodus perimeter for manganese as of synthetic provisions exposed resources is 0.6 mg/kg. Directive (98/83) specifies a perimeter of 50 μg/L in the EU drinking water as a industrial maximum value.

Nordic Interpretation

10% of COT EVM poisonous brink would consequence in 0.9 mg/kg.

Analysis

Around 0.02 mg/kg in food is the maximum value of observation through ICP-MS.

FCM

In the fabrication of steel, about 90% of the manganese is used. As well, in the production of pigments and glass manganese is used.

Raw Food

Vegetables, cereals, nuts plus some fruits have the uppermost confines of manganese.

Mercury (Hg) – Impurity

What and Where

From volcanic in addition to industrial activity primarily mercury contamination originates. Mercury gathers in the food chain with the uppermost concentrations found in large rapacious fish like tuna, marine mammals as well as swordfish. As well, fresh water fish might have elevated concentrations in specific regions. Characteristic dietetic supplements range as of 0.02 to 0.13 μg/kg bw per day.

Toxicological Sound Effects

The rising head of fetuses as well as little offspring organs are most sensitive. Mercury is poisonous to nervous structure as well as poses a danger on behalf of slow mind growth, development as well as wisdom in favor of offspring. One of the most hazardous elements for humans is methyl mercury.

Risk Assessments

In 2012 EFSA has put a bearable weekly ingestion (TWI) designed for methyl mercury 1.3 μg/kg bw per week.

Europe Council Limit Value

In 2013 the Europe council has put a definite discharge maximum value for mercury at 0.003 mg/kg allocate 10% of the toxic protection maximum value of JECFA of 0.004 mg/kg bw per week as well as assume that a 60 kg human being devour 1 kg of sustenance.

Further Limit Values

In guideline 1881/2006 EU utmost confines for mercury in fishery products as well as fish. The level alters among 0.5 as well 1.0 mg/kg. For meat, fruits and vegetables EU utmost levels are place in the rule 396/2005 for insect killer remains. Command (98/83) stipulates an intensity of 1.0 μg/L in the EU consumption fluid.

Nordic Remarks

No annotations.

Analysis

Almost 0.01 mg/kg in food is the limit of observation by means of ICP-MS.

FCM

In food contact substances mercury has no recognized exploit. Mercury outlines alloys, mixture, with about every additional metal apart from iron. Dental amalgam is one case in point, which comprise silver with tin suspended in mercury. Dental amalgam is banned in many countries because of its severe harmful effects.

Raw Food

In fishery products plus fish like methyl mercury, mercury is found mainly. In pigeons and ducks high confines of mercury have been seen in areas with a hot blemish contagion.

Processed Food

To the protein, tiny proportion mercury is bound. Fish oils for this rationale typically have little concentrations of mercury.

Molybdenum (Mo) – Main Component

What and Where

In many animals as well as plants molybdenum is present. Almost from 89 to 500 µg/person per day is typical dietetic ingestion.

Toxicological Effects

Development as well growth is affected by molybdenum.

Risk Assessments

In 2006 EFSA put a maximum bearable concentration of 0.6 mg per day for adults as well 0.1 mg per kg for children 1–3 years. In 2013 EFSA has concluded a sufficient ingestion, the concentration of 65 µg per day meant for adults.

Europe Council Limit Value

In 2013 the Europe Council put a particular discharge value of 0.12 mg/kg allotting 20% of the toxic safety concentration by EFSA of 0.6 mg per day as well presuming an individual of 60 kg utilizing 1 kg of sustenance apiece daily.

Further Limit Values

Legislation (98/83) specifies no value in the EU consumption fluid.

Nordic Remarks

TDI corresponds to 0.06 mg/kg, 10% of EFSA.

Analysis

Almost 0.006 mg/kg in food is the concentration of calculation through ICP-MS.

FCM

High-temperature steels as well high resistance alloys has molybdenum. Oxides of molybdenum are consumed in ceramics pigments.

Raw Food

In beans, peas, liver, spinach, lemons, tomatoes as well wheat germ molybdenum is present.

Nickel (Ni) – Main Component

What and Where

Nickel can be found in all soils and is emitted from volcanoes. Nickel is vastly used for alloys and in plating.

Typical dietary intakes range almost from 1.49 to 3.83 μg/kg bw/day.

Toxicological Effects

Around 10–15% of the population, frequently women, disturb with nickel allergy by nickel intake from food. These consumers get explode eczema by oral ingestion of still small concentrations of nickel. Kidneys, lungs, spleen, in addition to the myeloid system are the chief goal for the toxicity of verbally taken salts of nickel. Nickel is carcinogenic by breathing.

Risk Assessments

In 2008 and 2011 WHO put a TDI 0.012 mg/kg bw per day. In 2005 EFSA accomplished that endurable higher ingestion intensity could not be placed because of insufficient data in favor of dermal responses.

Europe Council Limit Value

In 2013 the Europe Council put a characteristic liberate perimeter of 0.14 mg/kg allotting 20% of TDI of 0.012 mg/kg bw per day placed in WHO in addition to assume a 60 kg individual overwhelming 1 kg of sustenance apiece day.

Supplementary Limit Values

Decree (98/83) stipulates a limit 20 μg/L like a practical limit in the EU drinking fluid. 2 mg/dm2 as of FCM is release level of Finland.

Nordic Interpretation

TDI communicates to 0.07 mg/kg, 10% of the WHO.

Analysis

Around 0.06 mg/kg in food is the perimeter of observation through ICP-MS.

FCM

The major exploitation of nickel is in the manufacture of elevated value, decay opposing alloys with iron, aluminum, copper, molybdenum, zinc as well as cadmium. Process tackles, milk, wine tanks and cook ware are stain less steel containing nickel. Nickel plated stuff are fewer tough as well as decay opposing than stain less steel in addition to then not simply utilized for editorials in exposure to foodstuff.

In earthenware glasses as well as for enamel frits nickel oxide is consumed. In glazes, in addition to for coloring ceramic objects, nickel carbonate is utilized.

Raw Food

In every category of sustenance, nickel is present in minute amounts, but in grains, cocoa, seeds, and nuts at higher levels.

Recommendation

For direct exposure to provisions nickel plated substance is not appropriate.

Silver (Ag) – Main Component

What and Where

Silver is utilized in favor of dental fillings as well as salts of silver may be consumed for e.g. disinfection of consumption fluid. As well, like complexion mediator in favor of confectionery decorations as well as into alcoholic beverages silver is consumed. 1.29 to 3.47 μg/kg bw per day are specific nutritional ingestion range.

Toxicological Effects

Anemia, growth retardation, cardiac swelling as well as degenerative deviation of the liver may result in repeated introduction to silver.

Risk Assessments

In 2005 the European commission put a collection of constraints in favor of exodus in favor of material have silver on 0.05 mg/kg foodstuff based on a WHO quantity of 0.39 mg/person per day since 2008 as well as 2011. Food exposed resources contain nano level silver have not been measured moreover ought to be expected independently on a case by case foundation.

Europe Council Limit Value

In 2013 the Europe council has put a special discharge concentration of 0.08 mg/kg. From 2011 this limit is based on ingestion information as of ANSES of 1.29 μg/kg bw per day as well as presuming an individual of 60 kg utilizes 1 kg of provisions every day.

Supplementary Limit Values

As a provision enhancer silver is utilized. Order (98/83) stipulates no definite perimeter in the EU drinking fluid.

Nordic Remarks

No comments.

Analysis

Almost 0.08 mg/kg in foodstuff is the concentration of calculation by means of ICP-MS.

FCM

In the manufacturing of cutlery as well as table ware silver is utilized. Silver alloys with 7.5% copper are recognized like excellent silver.

Processed Food

For confectionery decorations as well as alcoholic beverages it is useful in coloring agents.

Thallium (Tl) – Impurity

What and Where

Thallium is found as an impurity in alloys as well as in iron, lead, zinc copper ores. Complexes having thallium are odorless, colorless as well tasteless. In some places of the world salts of thallium are utilized in insect plus rodent poisons. 0.01 μg /kg bw per day to 0.02 μg /kg bw per day are specific nutritional drinking and eating ranges.

Toxicological Effects

People who consumed high concentrations of thallium in a small period of time studies have reported diarrhea, vomiting, effects on the nervous system, liver, lungs, kidneys, and heart as well temporary hair loss.

Risk Assessments

In 2009, 0.01 μg/kg bw an RFD has been set by the US EPA.

Europe Council Limit Value

In 2013 the Europe Council put a definite discharge concentration of 0.0001 mg/kg allowing 10% of the toxic suggested dosage as well presuming a 60 kg individual overwhelming 1 kg of provisions.

Other Limit Values

For the quantity of thallium in foodstuff exposed equipment, French commandment put an utmost concentration of 0.05%. Command (98/83) stipulate no maximum value in the EU drinking fluid.

Nordic Remarks

No comments.

Analysis

Almost 0.0002 mg/kg in foodstuff is the maximum value of calculation by means of ICP-MS.

FCM

In the pharmaceutical industry, the electronics industry as well as in glass production thallium is used.

Raw Food

In potatoes, fish, as well as vegetable thallium, is present at standards around 0.001 mg/kg.

Tin (Sn) – Main Component

What and Where

Tinplate is utilized in steel canisters, as of somewhere it is able to roam slowly in the foodstuff. For a few kinds of sustenance by means of a light color, e.g. pineapple and asparagus, it works as an antioxidant. 23 to 82 μg/kg bw per day is the distinctive food ingestion range.

Toxicological Effects

Vomiting and diarrhea caused by ingestion of tin. The result depends upon the quantity of tin in the provisions quite on the ingestion per kg of body weight.

Risk Assessments

14 mg/kg bw/week tolerable weekly intake put by JECFA. JECFA has confirmed that 250 mg/kg in other preserved kinds of provisions can pose severe sound effects concentrations of as well as 150 mg/kg in canned beverages.

Council of Europe Limit Value

The Council of Europe has in 2013 set a specific release value of 100 mg/kg in approximation with the EU maximum limits in Regulation 1881/2006.

Other Limit Values

In EU law 1881/2006 for tin canned food concentrations has been set. The quantities are 200 mg/kg for preserved provisions additional than beverages; 100 mg/kg meant for canned beverages in addition to 50 mg/kg for container victuals for babies, small children along with nutritional provisions for distinctive remedial purposes for infants. Order (98/83) stipulates no limit in the EU drinking water.

Nordic Remarks

No comments.

Analysis

Approximately 0.003 mg/kg in food is the quantity of calculation via ICP-MS.

GMP

Codex Alimentarius has prepared a policy of practice for avoidance plus minimization of inorganic tin in preserved food.

Process Conditions: High temperatures and oxygen may progress the discharge of tin. Containers should be cooled rapidly to 35–40 degrees. The discharge of tin is caused by the shelf life of the canned foodstuff. The airspace over the food within the can ought to be reduced. Sealing of the container with vacuum be able to diminish the amounts of oxygen and thus also the discharge of tin.

Storage: The speed of release will double for each 10 degrees increase in storage warmth. So, low storage temperatures are preferred.

Choice of Materials: Yield must observe with the typical stipulation, e.g. ISO and ASTM.

FCM

In steel containers, tinplate is used.

Raw Food

Particularly vegetables, as well as fruit through a low pH, are able to boost the discharge of tin.

Processed Food

A low feature of tin plated containers for e.g. pineapple may be a problem. Low corrosion accelerators and pH, e.g. nitrate, boost the discharge of tin.

Recommendation

Do not accumulate foodstuff in open containers, as opening the can will smash the varnish sealing. This poses direct contact between metals and food in the canister. In addition, oxygen will develop the release of tin as of the container.

Titanium (Ti) – Main Component

What and Where

In the earth's crust titanium is the ninth most plentiful constituent. Titanium is utilized like a color additive in a wide range of foodstuffs, e.g. pharmaceuticals, inks, toothpaste, dairy products, soft drinks as well as confectionery. 0.3 to 1 mg/person per day is the specific dietetic ingestion range.

Toxicological Effects

Titanium looks to be basically inert, i.e. it will cross the human body. Though, it is appraised that 0.8 mg per day will be captivated.

Risk Assessments

In 1969 JECFA did not place an ADI for titanium oxide owing to be deficient in of vital fascination as well as storage. In 2005 EFSA agreed with this.

Europe Council Limit Value

In 2013 the Europe Council has determined not to place a specific liberate concentration as titanium is measured to be basically inert.

Other Limit Values

Decree (98/83) explains no particular perimeter in the EU drinking water.

Nordic Remarks

No comments.

Analysis

No explanation.

FCM

Stain less steel is able to have up to 1% titanium. In white stains in e.g. lacquers, paper, enamels, paints coatings also plastics titanium oxide is used. As a food additive titanium oxide is used.

Vanadium (V) – Main Component

What and Where

In pharmaceuticals and in earthenware dyes vanadium is used. Above 80% of the world's manufacture of vanadium is used in steel for tackles plus surgery apparatus in addition to in alloys like ferrovanadium. It has a fine rust opposition against hydrochloric acid, sulphuric acid, alkalis along with salt water. Almost from 6 to 156 μg/person per day is the typical nutritional intakes assortment.

Toxicological Effects

On lungs, spleen, kidneys along with blood pressure vanadium leave harsh sound effects. In rats, developmental toxicity has been measured.

Risk Assessments

In 2004 and 2009 EFSA found so as to an experience of 0.01 to 0.02 mg/day is at least three instructions of amount less than the dosage that causes gastrointestinal sound effects in body builders taking vanadium as enhancements.

Europe Council Limit Value

In 2013 the Europe council put a definite discharge concentration of 0.01 mg/kg using the concentration of 0.01 mg per day place by EFSA as well as assuming a 60 kg person overwhelming 1 kg of foodstuff per day.

Other Limit Values

Dictate (98/83) stipulate no value in the EU drinking water.

Nordic Comments

No comments.

Analysis

About 0.02 mg/kg in food is the concentration of calculation as a result of ICP-MS.

FCM

Vanadium alloys may be utilized to generate special gears e.g. adding around 1% of vanadium to steel forms an extremely shock-resistant alloy. Vanadium steel is used for e.g. knives, spare parts and axes for rotating equipments. In glass pigments with blue along with green colors and ceramics vanadium is present.

Raw Food

Mostly at hand in mushrooms and seafood, but as well in a lot of vegetables and fruits in extremely low concentrations.

Zinc (Zn) – Main Component

What and Where

Zinc is present in many kinds of foodstuff and beverages and is a vital trace element. 8 to 16 mg/person/day are distinctive nutritional ingestion ranges.

Toxicological Effects

Zinc is not stored in the stiff and surplus ingestions upshot in reduced absorption and enhanced excretion. Nonetheless, cases of zinc toxic effects have been accounted, e.g. from galvanized iron cans asserting acidic drinks. Symptoms of severe sound effects contain diarrhea, abdominal cramps, epigastric pain, vomiting, and nausea, while chronic toxicity is connected to improved copper emission and reduced copper maintenance.

Risk Assessments

In 2003 the scientific committee on food put an upper concentration for zinc at 25 mg per day for adults as well as 7 mg per day for children aged 1–3 years, based on sound effects on the copper balance as a vital endpoint. In 2014 EFSA accomplished that zinc gives to typical role of the immune scheme.

Europe Council Limit Value

In 2013 the Europe council places a definite discharge amount of 5 mg/kg allotting 20% of the toxic threshold from EFSA at 25 mg per day as well as presuming a 60 kg individual overwhelming 1 kg of foodstuff every day.

Other Limit Values

According to directive 10/2011, 25 mg/kg is the specific movement value for synthetic food exposed materials.

Directive (98/83) specifies no limit in the EU drinking water.

Nordic Comments

10% of the EFSA higher acceptable level corresponds to 2.5 mg/kg.

Analysis

Around 0.1 mg/kg in food is the concentration of calculation by means of ICP-MS.

FCM

Mainly zinc is significant in galvanized iron. Zinc protects iron from corrosion, as it is a stronger reducing agent. Furthermore, zinc is used in alloys like bronze and brass. Zinc coated steels are consumed in silos for storing foodstuff. Zinc oxide and zinc sulphide are used in ceramics, paints and other products. Zinc may consist of small concentrations of lead and cadmium as impurities.

Raw Food

The chief suppliers to zinc ingestion are whole grain cereals, meat and milk products such as cheese. Moreover, peanuts and oysters surround high amount of zinc.

Processed Food

Cheese is the major source.

Recommendation

Do not use galvanized iron to expose with moist otherwise liquid food.

SUMMARY

Alloys and Metals are immensely applied as victuals contact supplies, e.g. as household utensils and as practice paraphernalia in the foodstuff production. So, they are a possible source of food contagion. Substances migration from food exposed materials to provisions should not take place in concentrations that jeopardize human fitness. Related for food exposed resources made from alloys and metals are the exodus of metals, both the key components as well as foreseen contaminations.

Domestic manage stand on an affirmation of conformity, DoC, and sustaining records at the manufacturers in addition to importers are major fundamentals to control this contagion and to make sure conformity with the set of laws. This is measured as a common constituent of feature guarantee, even though the European law does not particularly need a DoC for alloys and metals used as provisions exposed supplies.

CHAPTER

10

PREVENTIONS FROM HAZARDS

GENERAL DESCRIPTION

It is hard to find anybody who has not encountered an objectionable moment of foodborne illness at least just the once in the past year as well as food protection is everybody's concern. By microbial pathogens, foodborne illnesses might outcome as of the utilization of foodstuff infected, radioactive materials otherwise toxic chemicals. Another rising disruption is food allergy. Though numerous foodborne ailments might be self-limiting, various are able to be extremely severe as well as even pose bereavement. In the circumstance of varying foodstuff behaviors, make sure provisions protection is appropriate gradually more significant, the globalization of our foodstuff contribution as well as popularization of gathering cookery establishment. The stipulate to reinforce foodstuff protection organizations within as well as among the entire states are becoming supplementary in addition to more obvious, as the provisions supply becomes gradually more globalized.

The "WHO Five keys to safer food" provide the same as the origin for learning programmes to train provisions handlers as well as inform the customers. In averting foodborne infirmity, they are especially important.

The five keys are the same as subsequent:

- Wash each and every one utensil, platters, plates as well as crockery immediately when used. Keep foodstuff surfaces fresh.
- Separate cooked foodstuff as of raw foodstuff.
- To the appropriate heat, cook groceries properly.
- Both for storage as well as serving keep foodstuff at protected heat.
- Utilize protected uncooked resources as well as water.

Although, as compared to normally offered foodstuff organic foodstuffs are costly, there is ability amongst fitness-aware clients to eat less, but procure organic provisions.

During yield, processing, transport, storage and finally during foodstuff preparation as well as storage by means of clients, ensuring food safety requirements appropriate awareness. Processed, ready-to-eat otherwise frozen foodstuff is gaining attractiveness in recent years because of varying provisions behaviors, diversification of product, and hectic way of life as well as accumulation manufacture practices.

FOOD SAFETY AND FOOD FEATURES

Food Features

Features include negative as well as positive features so as to manipulate a product's significance to punters. Negative characteristics might be observable contamination with dirt, spoilage, staining, and otherwise bad tastes or odors as positive attributes so as to reveal superior value might be the color, origin, texture, and flavor as well as handing out methods of the foodstuff. This difference in quality, as well as safety, has suggestions meant for civic strategies in addition to impacts the character as well as substances of the foodstuff manage scheme mainly suitable together pre arranged general purposes.

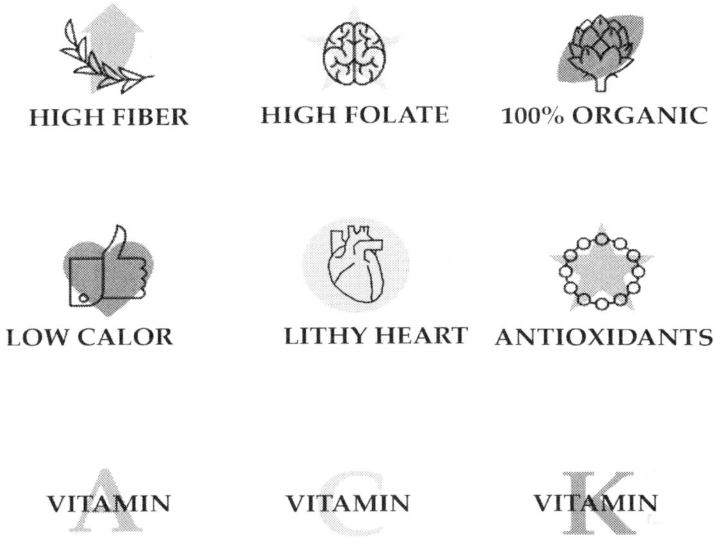

Figure 10.1: Food Features.

https://thumb1.shutterstock.com/display_pic_with_logo/2381402/228332023/stock-vector-vector-modern-creative-line-icons-on-healthy-organic-food-benefits-set-of-nine-food-features-such-228332023.jpg

Food Features Evaluation

For sensory assessment, a careful sampling of the foodstuff is vital. It is not at all times probably to identify by means of sensory processes only the infectivity of foodstuff via veterinary drug residues, pesticides as well as tarnishing. Purpose estimation is through which comprises physicochemical, chemical, and physical as well as microbial methods of investigation. Chemical techniques include the evaluation of the nutritive significance of foodstuff prior to as well as subsequent to food preparation, as well as to recognize the foodstuffs of adulterants Furthermore decomposition in foodstuff. The main generally working purpose estimation is the extent of substantial characteristics by means of the utilization of utensils. Estimations of the manifestation as well as the volume of foodstuff are as well significant.

Food Safety

Foodstuff protection is regarding handling, producing, preparing as well as storing provisions in such a way like to avoid illness as well as contagion in the provisions manufecturer series, as well as to assist ensures so as to foodstuff features as well as integrity be preserved to encourage high-quality fitness.

Figure 10.2: Food Safety.

https://www.fda.gov/ucm/groups/fdagov-public/documents/image/ucm261350.jpg

ORGANIC FOOD OR LOCALLY PRODUCED FOODSTUFF

Organic as well as locally formed foodstuff might have ecological profits like by means of smaller amount fertilizers otherwise pesticides. It is necessary in favor of distributors as well as farmers in the direction of utilize fine hygienic practices to reduce victuals infectivity. Customers must constantly cook provisions suitable as well as prepare, no matter wherever it is from.

Figure 10.3: Organic Food.

http://marianovikova.ru/wp-content/uploads/2014/12/Ovoshchi.jpg

STREET PROVISIONS

Street foodstuff wholesalers are an indigenous as well as traditional junk foodstuff forthcoming several states of the South-East Asia area. Street foodstuff wholesalers present enjoyable and cheap foodstuff to millions of customers. There are a lot of causative causes associated with the protection of street foodstuff; it is hard to say how protected street food is. As the majority of street foodstuff is served hot as well as cooked well, there is fewer threat of foodstuff poisoning. The hygienic conditions, water quality in addition to the intensity of sanitation, particularly in the summer time of year, might supply to experiences of foodstuff poisoning. However, as customers are also demanding featured foodstuff developments in sterile circumstances as well as substantial infrastructures are increasing.

Figure 10.4: Street Provisins.

https://i.ytimg.com/vi/m7BMd70vfoI/hqdefault.jpg

(HACCP) HAZARD ANALYSIS OF CRITICAL CONTROL POINT SYSTEM

The Hazard Analysis Critical Control Point system, or, HACCP, is developments organize organization so as to distinguish wherever risk may take place in the foodstuff formation method as well as sets into consign severe measures to avert the risks from happening. By severely controlling as well as monitoring every step of the method, there is not as much of chance for exposures to take place. HACCP is an essential fraction of current foodstuff manufacturing used to recognize as well as organize chief foodstuff hazards, for instance physical, microbiological as well as chemical contaminants. Such incorporated approaches facilitate enhanced customer security, efficiently encourage cultivation as well as the foodstuff handing out manufacturing, as well as encourages international as well as domestic foodstuff employment.

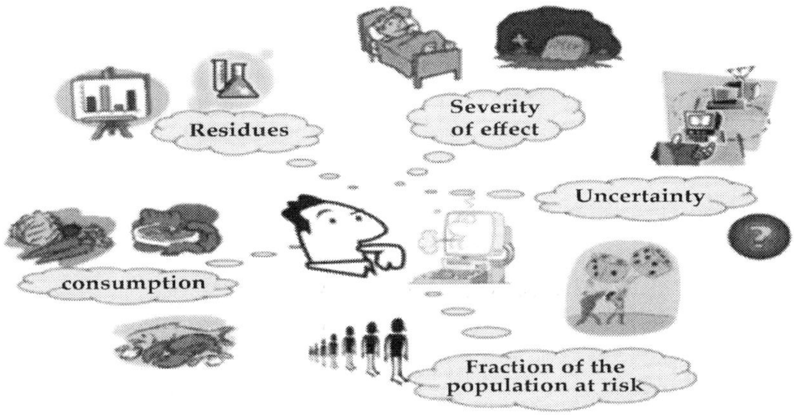

Figure 10.5: Hazard Analysis of Critical Control Point System.

http://www.safefoods.nl/upload_mm/c/2/1/50b27810-0fae-4f48-894c-ba203e55a526_120diverse20considerations1_1852f138_530x373.jpg

FOOD RECALL

During the organization of menace in reaction to foodstuff protection actions as well as crisis, food recall is an essential gizmo. A food recall is defined is an act in use to eliminate foodstuff that can pose a protection hazard as of the deal, sharing otherwise utilization by customers. In relation to food recall, the phrase "withdrawal" is exeplaind extensively.

A food recall might be started because of a complaint otherwise information as of an assortment of causes, producers, wholesalers, sellers, government organizations as well as users. Recalls are performed as a result of foodstuff productions to make sure so as to unsafe foods otherwise potentially dangerous are not devoted.

Foodstuff as well as the features of provisions yield are gradually more developed, progressed as well as used in a variety of places in the region of the globe. This globalization poses innovative confronts in the behavior of key actions associated to foodstuff recalls, like the trace-back in addition to trace-forward actions necessary in favor of a foodstuff confirmed or suspected to be risky.

PROCURE HARMLESS PROVISIONS IN THE MARKET

While Buying Fruits and Vegetables: Precautions

Be cautious because several vegetables as well as fruits might be synthetically decorated to provide the delusion of quality as well as freshness. The majority of clean fruits as well as vegetables preserve their cleanness for a little moment below perfect conditions of storage as well as it is the entire time superior to obtain vegetables in the evening or morning hours depending on harvesting sample. As the features are normally good as well as the value is small, so it is worthwhile to procure fruits as well as vegetables which are in season.

FRESH FISH RECOGNIZATION

Fresh fish has a stiff body, tight scales as well as firm flesh. Be confident to procure fish that have been stored on ice otherwise refrigerated and are not slimy or slippery to handle.

ASSOCIATED HEALTH RISKS: FISH PRESERVATIVES OR COLORANTS

The red color is a significant feature characteristic where carbon monoxide behavior of fish is explained. As a color change is utilized by consumers for the main estimation of feature, carbon monoxide treated fish has the ability to compose sub standard fish appear aesthetically further pleasant to customers.

The application of formalin to protect foodstuff is unlawful as it is a powerful toxin, as well as fish conserved by formalin is not suitable in favor of utilization.

FOOD SAFETY CONCERNS: SEAFOOD AND FISH

Fish as well as seafood can develop into unhygienic by microbes like Salmonella, Vibrio cholera, Shigella, Listeria, E. coli, by reason of individual's action otherwise underprivileged cleanliness as well as hygiene while foodstuff manufacturing as well as handing out. In addition, Methyl mercury is produced as a result of microbial action in marine surroundings from discarding of industrialized mercury as well as usual source of natural mercury. Definite parasite larvae might be found in fish that is why mainly fish ought to be cooked systematically.

FOOD LABELING: IMPORTANCE

Prepackaged foodstuff ought to be appropriately tagged in favor of lawful, administrative as well as technical causes. Food labeling may be some printed; graphic otherwise written matter, which occurs on the tag, accompany the foodstuff, otherwise is exhibited close to the foodstuff, counting so as to meant for the reason of encouraging its disposal otherwise sale. On the tag of prepackaged foodstuff, following data ought to appear:

- Food name.
- Ingredients list.
- Name and address
- Usage instructions

Label statement on packed foodstuff is extremely significant in favor of knowing the constituents as well as dietetic contents, in addition to assists in examination of the cleanness of the foodstuff as well as 'expiry' dates. It supplies evidence in favor of 'food withdrawal' otherwise 'food recall' if circumstances demands.

'USE-BY DATE' FOOD LABEL: IMPORTANCE FOR CONSUMER

'Use-by Date' moreover described as the suggested final expiration date otherwise consumption date is the ending of the predictable time after that the manufactured goods certainly will not have the eminence characteristics usually anticipated as a result of the customers provided that it is preserved less than the acknowledged situations for storage.

Examination of the 'use-by date' as well as 'best before' labels on foods is one of the primary thing customers ought to check prior to buying. It is a widespread perform used for super markets to decrease price of foodstuffs so as to will in a little while reach their "use-by-date" as a result particular attention ought to be compensated prior to buying such substance to evade buying foodstuff so as to might not be inspired prior to the ending date.

SAFE FOOD PREPARATION AT HOME

FOOD COOKING AND REHEATING

WATER SUITABLE FOR FOOD PREPARATION

In foodstuff manufacturing, water is significant. Before cooking, it is consumed to rinse fare, as well as utilized to perform as a food preparation intermediate, as well as it is consumed to rinse cans of provisions prior to as well as after production in addition to it is moreover consumed as the most significant beverage. For making ice, water consumed ought to as well be drinkable. The ice might be utilized to cool food otherwise added to colddrinks.

COOKING OIL USE AND REUSE FOR FRYING: AND HEALTH HAZARD OF REUSE OF COOKED OIL

Oil is typically added while the primary steps of food preparation as well as is mostly a significant fraction of cooking. It is advantageous to transfer a little oil inside the pan at the preliminary quite afterward while the saucepan is extremely warm.

The trouble with recycling oil is so as to it is able to generate 'free radicals' that may cause diseases in the long run. The free radicals might be carcinogenic that is able to pose melanoma in addition to as well atherosclerosis that may direct to enhance into dreadful cholesterol conentrations, overcrowding the blood vessels.

The subsequent actions must be measured to minimize fitness danger of reusing oil:

- Soybean, sunflower, canola as well as mustard oil encompass an elevated smoke point that is these don't break at elevated temperatures as well as are as a result suitable in favor of frying, counting profound frying wherever the provisions are submerged into the oil.

- Ensure the extra oil as of frying otherwise cooking is cooled down as well as then move into an sealed can by means of a filter.
- It is time to modify the oil, if recycled food preparation oil is dark in color otherwise is sticky or greasy.
- Uncontaminated the container by which food preparation is executed between utilization.

REDUCTION OF DIETETIC INGESTION OF PESTICIDES IN KITCHEN

It is hard for a customer to examine whether fruits otherwise vegetables have pest killer remains like mostly these don't have any obvious taste, smell otherwise illustration imperfection. The subsequent safety measures must be in use to minimize dietetic contact to insect repellent remains if farming foodstuffs be sold in general marketplace:

- Completely scrub as well as rinse vegetables and fruits. If possilble peel them.
- Throw back the large fish like the small ones include small duration to start as well as concentrate pesticides as well as further damaging remains.
- Spruce fat as of meat, fish as well as poultry; dispose of fats as well as oils in drippings as well as broths.

MEAT OR FISH: PROPERLY COOKING

It is essential to cook all animal protein appropriately to finish the bacteria that may pose foodborne ailments. Cooking times as well as temperature will fluctuate depends upon the kind of slice of animal protein as well as the method of food preparation.

While practicing animal protein at home, the subsequent instructions ought to be measured:

- Ensure there are no pink animal protein remains. Meat alters shade at what time it is safe to eat.
- When you cut the thickest ingredient of the animal protein by means of a skewer otherwise fork, the juices must sprint plain.

- If the animal protein is white as well as there is no pink tissue, chicken is finally safe to eat. Fish is completed while it is dense as well as peels off without problems with a fork.

Figure 10.6: Properely Cooked Fish.

http://www.molecularrecipes.com/wp-content/uploads/ingredients/xtransglutaminase-meat-glue-fish.jpg,q93afc3.pagespeed.ic.3K-I8mlInS.jpg

RAW FOOD SEPARATION FROM COOKED FOOD

Cross-contamination is the transport of damaging germs from one piece of foodstuff to a new by a non-food plane like equipment, utensils otherwise human hands. Impure water might be possible cause of foodstuff infectivity.

Figure 10.7: Raw Food Separation from Cooked Food.

http://www.homeanddecor.com.sg/sites/default/files/imagecache/hnd_revamp_1x1_large/blog/gallery_article/gallery_images/62110-raw-meat-1r.jpg

USE OF MICROWAVE OVENS

Several people have an intuition so as to microwave ovens are utilized particularly in favor of reheating.

Precautions: While Using Microwave Oven

Just make use of cookware so as to is particularly designed for utilization in the microwave oven. Just make use of foodstuffs so as to have been obviously tagged the same as protected to make use of in a microwave, whether prepared as of glass otherwise earthenware. A lot of plastics are not protected in microwaves, somewhere these may soften; make fractures so as to may because of leakage of foodstuff, otherwise melt, as a result, is more cautious. Most metals, counting metal paints on glass otherwise earthenware, are typically not protected both due to the utmost hazard of fire as well as sparks.

For microwave utilization covers the dish with a cover otherwise plastic envelopes accepted. Vent or loosen the wrap or lid to permit steam to emit.

Food Cooked in Microwave Oven

During microwave cooking, bacteria will be damaged only like in supplementary categories of ovens, so foodstuff is protected safe to eat in a microwave oven, if it is passionate to the essential temperature in general over 70 °C. In other words, microwave ovens are able to cook unequally as well as depart "cold spots" wherever damaging microbes is able to stay alive.

Frozen Food in a Microwave Oven

Before defrosting, remove foodstuff from its wrapping. Don't make use of plastic wraps as well as foam trays as these are not heat constant on elevated temperatures; warping otherwise melting of the wraps or else trays as of the warmth of the foodstuff might pose dangerous chemicals to percolate in. Microwaves are wrapped up preferentially through water sooner than ice, in addition to via the first coating so as to be able to soak up the rays.

Food preparation entire, swollen poultry in a microwave oven is not recommended, as foodstuff in a microwave oven be able to cook earlier on the exterior than foodstuff on the inside, the filling may not have sufficient moment to arrive at the warmth required to demolish damaging microbes.

STANDING TIME

Microwaves pose fat, water; as well as sugar molecules to vibrate 2.5 million times apiece second, to produce heat. After the oven is off otherwise foodstuff is detached as of the oven, the molecules carry on to produce warmth as they approach to a decline. This extra food preparation after microwaving discontinues is known as "carry-over cooking time," "resting time," otherwise "standing time." as well, microwave heating is typically rough with central mark of strong warmth might be close to areas of cool particularly if initially cold, as a result standing time permits for more still heat dispersal. In favor of so as to cause, instructions typically direct to allow a foodstuff "rest" in favor of a little record after rotating off the oven otherwise eliminating foodstuff as of the oven.

PASTEURIZATION: PASTEURIZED MILK

Pasteurization is the main accepted technique of heat action. It occupies heating milk to elevated sufficient temperature to exterminate dangerous microbes so as to be able to pose disease. Milk is frenzied to the elevated heat as well as afterward quickly refrigerated so as to temperature established in addition to temperature labile pathogens are killed. So, pasteurization only doesn't provide a protected shelf-stable item for consumption devoid of appropriate storage.

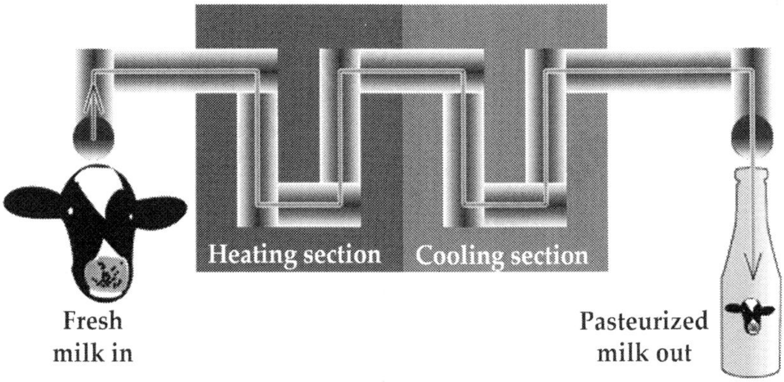

Figure 10.8: Pasteurization of Milk.

http://cdn4.explainthatstuff.com/how-pasteurization-works.png

TETRA PACK: MILK OR JUICE

The fruit juice, cream otherwise milk is pasteurized at elevated hotness (132 °C.), other than meant for a shorter moment that is called 'ultra-high temperature (UHT)' for foodstuff processing. Tetra pack foodstuff is safe as well as sterile to devour in anticipation of they're unlocked. After opportunity it ought to be accumulated in the refrigerator.

Figure 10.9: Tetrapack Milk.

http://www.afternoondc.in/userfiles/images/shutterstock_19025491.jpg

FOOD PRESERVATION AND STORAGE

Eventually, just ice-cold foodstuff will blemish. The cool hotness slows down microbial growth however they don't impede the expansion entirely.

It ought to be distinguished so as to several microbes similar to Listeria be able to raise fine in cross-contamination as well as refrigerator may happen if raw as well as safe to eat foodstuff otherwise salads are reserved in addition to open collectively. Cake, eggs as well as several foodstuffs might sop up taste as of supplementary foodstuff.

Figure 10.10: Food Preservation and Storage.

http://beforeitsnews.com/contributor/upload/155734/images/food%20preservation.jpg

PRECAUTIONS WHILE USING REFRIGERATOR OR FREEZER

All the time refrigerators ought to be kept dry as well as clean. The subsequent instructions will help to make sure provisions protection, quality as well as integrity of victuals:

- Don't overstock refrigerator. Overloading will decrease the movement of cold atmosphere in addition to compose hard the appropriate chilling otherwise cooling of beverages as well as foodstuffs.

- Poultry, raw meat in addition to seafood must be in closed wrapped otherwise containers strongly to stop raw juices as of contaminating other provisions.
- Wrap leafy vegetables or cover food with paper to keep wetness as well as avoid them as of picking up odors as of supplementary foodstuff.
- One time a week, make it a practice to pitch out unpreserved foodstuff so as to must no longer be eaten.

Provisions conservation is considered to avert decay as well as spoilage of recurring as well as unpreserved foodstuff by raising the shelf-life of foodstuff therefore escalating the contribute in addition to accessibility the whole year.

FOOD PRESERVATION: COMMON METHODS OF PRESERVATION

The fundamental standard of every conservation procedure, whether carried out in the home otherwise commercially, is to confine foodstuff spoilage so as to foodstuff be able to used carefully in a edible form at a later point. This is attaining by warmth, drying, chemicals, radiation, fermentation as well as cold. To make use of ice or else snow to conserve foodstuff was recognized to early man. Fermentation of undeveloped make moreover was recognized as of early times as well as it is still experienced in favor of conservation of vegetables like gundruk, kimchi, sinki as well as pickles.

In microbiological conditions there are mainly two methods of customary food conservation:

- 'Altering environmental conditions' by which spoilage microbes are not capable to raise in the foodstuff, as in pickling, freezing, dehydration, smoking, salting etc. Such circumstances may be formed by means of elimination of water, use of oil, use of acid as well as spices, utilization of chemical preservatives Moreover use of minimum temperatures.

Figure 10.11: Food Preservation.

http://maximumsurvival.net/wp-content/uploads/food-preservation.jpg

- 'Elimination of enzymes as well as microbes' in which most of the microbes present in the foodstuff are killed in addition to enzymes are inactivated, example, in cooking, canning, blanching, irradiation etc. High-fat foodstuff doesn't accumulate fine in the icy condition. Oily fish are the mainly appropriate for canning. Canning keeps the normal taste of the fish. Botulinum spores are not all the time killed by the elevated temperature of the canning procedure, however the toxin so as to these microbes formulate does not take place in acidic circumstances. So, pickling otherwise canning of vegetables otherwise meat in sealed containers where possible must have sufficient sour to evade poison development.

MICROBES INSIDE REFRIGERATOR: MICROBIAL GROWTH PREVENTION

Fungi in addition to bacteria might rise in the refrigerator if cleanliness as well as dryness is not preserved. Eventually, spoilage microbes are able to raise at low temperature as well as they pose foodstuff to extend off or else smells and bad tastes, wholesomeness as well as compromising quality of food. Pathogenic microorganisms don't generally influence the smell, taste, or else manifestation of a foodstuff, other than there is a possibility of cross-contamination if cooked as well as raw foods are kept together.

REFRIGERATOR CUT

To preserve internal warmth protected stay the door closed in so far as feasible. Provisions inside be protected as long as power is out no over 4 hours, with no opening the doors. Forever dispose of any substance in the refrigerator that has come into contact with raw meat juices.

PREVENTIONS FROM STRONG FOOD ODORS INSIDE REFRIGERATOR DUE TO FOOD SPOILAGE

After power outages strong provisions odors are frequent. The subsequent exploit ought to be in use to get rid of undesirable, unpleasant foodstuff odors;

- The freezer otherwise refrigerator must be unplugged as well as empty while cleaning.
- Remove all detachable fractions as well as clean by means of mild soap as well as water.
- Utilize vinegar solutions, one cup vinegar per gallon of water to clean the inner walls of the freezer or refrigerator.
- If the smell persists, allow place two to three days, altering vinegar every eight hours.

CANNED OR PACKAGED FOODS

There is no time concentration on how extensive unopened containers will be protected. The common suggestion for preserved foodstuff with low acidity like meats as well as vegetables is so as to they will be high-quality for 2 to 4 years. Acidic canned foodstuff like tomatoes, fruits as well as fruit juices will typically even be fine value for 1 to 2 years.

Certainly long storage will alter the feature of the foodstuff before the protection is exaggerated. Jars as well as cans with stuffed sides otherwise tops must also be thrown away; don't eat any of the foodstuffs like it might be impure with botulinum poison that might be critical if utilized. Unopened bottles otherwise jars with visible discoloration or bubbles, ought to also be thrown away.

Figure 10.12: Canned or Packaged Food.

http://occ144datkn3vrjlq7r63p19.wpengine.netdna-cdn.com/wp-content/uploads/2015/03/food-for-thought-books-02.jpg

It is forever a fine scheme to turn packaged or canned foodstuff, as well as to place dates on packages so that one be able to recognize how old the substances are.

PLASTICS USED FOR FOOD HANDLING AND STORAGE

Plastic packaging plays a significant responsibility in the shelf life as well as easiness of cooking as well as storage for several foodstuffs as well as the majority are protected to utilize provided so as to they are utilized properly. Several categories of plastics resources that are vastly consumed for treatment as well as food storage as well as water are subsequent:

Polyethylene Terephthalate (PET)

It is utilized to form water, soft drink, ketchup, sports drink, as well as peanut butter, salad dressing bottles, jelly, jam jars in addition to pickle. It is heat resistant, strong as well as resistant to acidic foods plus gases. It may be opaque otherwise transparent.

High-Density Polyethylene (HDPE)

It is consumed to produce water, milk as well as yogurt, juice bottles along with grocery moreover margarine tubs, retail bags also trash. High-density polyethylene is strong as well as stiff however is not stable to heat, that is it melts at a comparatively low temperature.

Low-Density Polyethylene (LDPE)

It is consumed to produce some bread, films of various sorts as well as squeezable bottles in addition to frozen food bags. It is comparatively translucent.

Polypropylene (PP)

It is supplementary harder; denser, high temperature resistant as well as extra clear than polyethylene thus is utilized for heat-resistant microwavable wrapping as well as salad otherwise sauce dressing bottles.

Polycarbonate

It is heat resistant, clear as well as hard-wearing in addition to mostly utilized to compose sterilizable baby bottles and refillable water bottles, plastic coating for metal cans, microwave ovenware. At elevated concentrations of introduction, bisphenol A is potentially dangerous as it impressionist the female hormone estrogen.

As well, polyvinyl chloride (PVC), as well as polystyrene (PS), are as well utilized while foodstuff handling as well as transportation in supermarkets.

Reduction of the Migration of Chemicals from Plastic into Food

All synthetic is prepared as of chemicals so as to have the potential to damage health of a person. Subsequent these straightforward instructions will assist to decrease the exodus of chemicals as of plastic into foodstuff:

- Follow instructions of manufacturers while utilizing domestic plastics like bags in addition to cling films.

- For the responsibility, use the accurate category of synthetic, for example, just utilizes microwave-safe synthetics in the microwave.
- Don't let cling film touch the foodstuff through microwave food preparation like it melts by the side of a low temperature. In a lot of cases, the film is supposed to be disinterested prior to food preparation in a microwave.
- Reuse synthetic containers so as to be foodstuff companionable, in the technique the unique foodstuff was obtainable. For illustration, you can chill foodstuff in ice-cream containers however don't warmth them in the microwave, they were considered in favor of utilization on frozen foodstuff.

FOOD ADDITIVES AND FORTIFICATION

Food Additives

Foods additives are the materials usually not used as a foodstuff itself as well as are not generally consumed as a characteristic constituent of the foodstuff, whether otherwise not it has nutritive limit. The word doesn't contain impurities, otherwise materials added in foodstuff meant for improving otherwise preserving otherwise dietetic behavior, or else sodium chloride.

Figure 10.13: Food Additives.

http://3.bp.blogspot.com/-4GbvXwz1jCw/TwGLd0T3aZI/AAAAAAAAAxU/LQ-Dr0Y441SA/s1600/otrovi-u-hrani.jpg

Food Fortification

Food fortification is expressed as the addition of one otherwise more than one crucial essentials to a foodstuff. Food fortification has the benefit of being capable to distribute nutrients to huge pieces of the populace devoid of having need of fundamental alternations in foodstuff utilization samples.

The speedily varying style of life as well as rising confidence on supplementary high development foodstuff has been utilized to substantiate the counting of essentials to a growing variety of foodstuff in array to make sure dietetic sufficiency of the food. Provisions machinery has been urbanized in favor of micronutrients as well as vitamins fortification in milk as well as milk foodstuffs, cereals as well as cereal based foodstuffs, oils as well as fats.

Food Fortification Main Concern Areas

In 2000, the World Health Report, recognized iron, iodine, zinc as well as vitamin A lackings because amongst the mainly severe fitness hazard factors of the world. The center of the international community meant for support of fortification of foodstuff has until now on the three most common shortages: vitamin A, vitamin B complex, iron as well as iodine.

FOOD ADULTERATION AND CONTAMINATION

Food Adulteration

In states of the South-East Asia area food adulteration is regrettably ordinary consign and is an immoral in addition to mostly unlawful mismanagement. Food adulteration is generally observed in its most rudimentary variety wherever illegal materials are both used as well as added to wholly otherwise partly or else alternate fit constituent otherwise to synthetically generate the intuition of cleanness in unhygienic foodstuff. Ultimately, this adulteration causes the customer to be whichever embittered economically otherwise of poorer quality seeing that a sufferer of disease or illness. Though, sufficient safety measures in use by customers at the instance of acquire of these foodstuffs may permit them to evade procuring these adulterated foodstuffs.

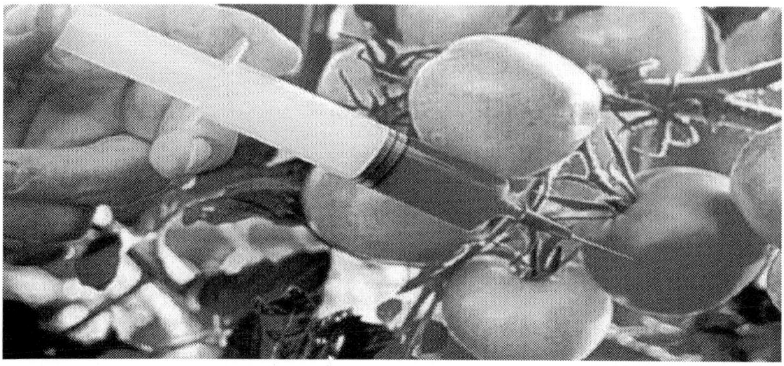

Figure 10.14: Food Adulteration.

https://3.bp.blogspot.com/-LBw4qlxa_VY/VmiF67bfXDI/AAAAAAAAADE/EbFSbqCgEqo/s1600/Like%2Bmost%2Bfood%2Badulteration%252C%2Bthreats%2Bto%2Bpublic%2Bhealth.jpg

Following are the several instances of adulteration of foodstuff:

- Chili fine particles: Residues of brick, talc or saline fine particles, residues
- All flavors: Dust, dirt, powder, other kernels
- Honey: Syrup of sugar, jaggery
- Beverages and ice cream: Saccharin

Following are several patterns of artificial dyes:

- Emerald chili, sour gourd, green peas, green vegetables: Malachite green
- Chili residues: Rhodamine B
- Turmeric fine particles: Lead chromate

Food Contamination: Sources of Contagion

Food contaminant is any substances unintentionally added to foodstuff that may found in the food because of the formation having procedures done in animal husbandry, crop yielding in addition to aquaculture, production, dealing out, grounding, handling, stuffing, wrapping, transfer otherwise investment of these foodstuff or else as a consequence of ecological contagion. During handing out by reason of outflow of mineral oil within manufacturer stroke, edible oil might be unhygienic.

Figure 10.15: Food Contamination.

http://coachwithlisa.com/wp-content/uploads/2015/01/canstockphoto8171126-471x508.jpg

In states of the South-East Asia country, food contagion is still a larger hazard as a result of the deficiency of manufacture crop standards, foodstuff conducting principles, as well as ecological rules.

PESTICIDES' EXPOSURE REDUCTION

It is usually recognized so as to insecticides or pesticides as well as fungicides or herbicides are consumed in favor of insects, pest and fungal and weed manage. These agrochemicals are accessible more the offset; farmers might consume these chemicals devoid of suitable indulgent otherwise management. In view of, the potential fitness exposure of dietetic ingestion of remains of these substances, the acceptable maximum dietary intake (ADI) as well as maximum residual limit (MRL) on the source of proof based data as well as hazard estimation has been advised in the Joint FAO/WHO Meetings on Pesticide residues provides free systematic professional suggestion on insect repellent remains, as well as their residual concentrations.

MAXIMUM RESIDUAL LIMIT (MRL) AND ACCEPTABLE DAILY INTAKE (ADI)

Acceptable Daily Intake (ADI) of a chemical is the every day ingestion which, while an intact life span, emerges to be devoid of considerable danger to the fitness of the customer. It is articulated in milligrams of the compound per kilogram of body weight.

Maximum Residual Limit (MRL) is the highest amount of a pesticide remains uttered as mg/kg, suggested by means of the Codex Alimentarius Commission to be officially acceptable in foodstuff supplies as well as living thing feeds. MRLs which are mainly anticipated to be appropriate in worldwide dealings are derivated from estimation through the JMPR following both:

- Toxicological estimation of the pesticide as well as its remains
- Appraisal of remainder information as of supervised assessments as well as administer uses counting those dazzling nationwide foodstuff undeveloped practices.

ADI and MRLs are in addition indomitable for veterinary drug remains as well as antimicrobial materials as these compounds are utilized for development encouragement as well as animal illness avoidance as well as management.

FOOD RELATED HEALTH RISKS AND VULNERABILITIES

FOOD NATURAL TOXINS

Several foodstuffs have a toxin in their uncooked form so as to be made a smaller deadly one time cooked. Red kidney beans are toxic when used in raw form because these may include a high quantity of lectin. Severe stomach ache, diarrhea in addition to vomiting could be caused by as few as four or five raw beans. Alike, in the raw state, the castor bean lectin ricin is tarnished in favor of posing deaths of offsprings as consumed.

Figure 10.16: Natural Food Toxins.

http://mindaberbeco.scienceblog.com/files/2013/09/mushroom2.jpg

A glycoalkaloid 'solanine', present in elevated maximum amounts in green areas otherwise budding of the potato, is intensely poisonous to individuals. Occasionally cucumber might have a cluster of ordinary poison called 'cucurbitacins'. These poisons provide cucumber a sour flavor.

Cabbage, as well as associated vegetables, contain thioglucosides that might be captivated in individuals by low nutritional iodine, as well as might add to thyroid swelling.

Cassava should be appropriately prepared before eating, to prevent introduction to these poisons. Peelings as well as slice, the cassava also then cook it methodically, either by means of boiling, roasting otherwise baking.

The majority of the natural toxins present in fish are formed as a result of types of unsurprisingly taking place aquatic algae. Ciguatera fish toxicity is differentiated by tingling in addition to numbness of the lips as well as tongue, diarrhea, vomiting Moreover is related the utilization of toxin-contaminated tropical plus subtropical reef fish. Unfortunately by normal processing or cooking these poisons are not damaged.

Wild Mushrooms Toxin: Precautions

In the countries of South-East Asia and South Asia, foodstuff toxicity as of the utilization of venomous feral thrives has been accounted recurrently in heavy rain period. A lot of poisonous mushrooms are moderately alike in look to their reduced amount of poisonous cousins, consequently over and over again simply to error.

The bulk of lethal poisoning of mushroom happens because of intake of the 'death cap', Amanita Phalloides, because of the high concentrations of, a potent cytotoxin, 'Amatoxin'. Single mushroom has sufficient venomous to execute an adult. The start of signs takes place within 6-24 hours otherwise further after intake of thrives. Lethal poisoning is typically related to the late start of signs that are extremely relentless, the poisonous result on the kidney, liver as well as nervous system. In favor of complete protection prevent every feral mushroom, except certainly identified like non-poisonous.

Mycotoxins

Mycotoxins are the cluster of physically taking place chemical substances formed by means of definite fungi or moulds. These may be manufactured by a number of fungi within products as well as nourish through manufacture, transportation, storage, frequently underneath humid as well as warm circumstances.

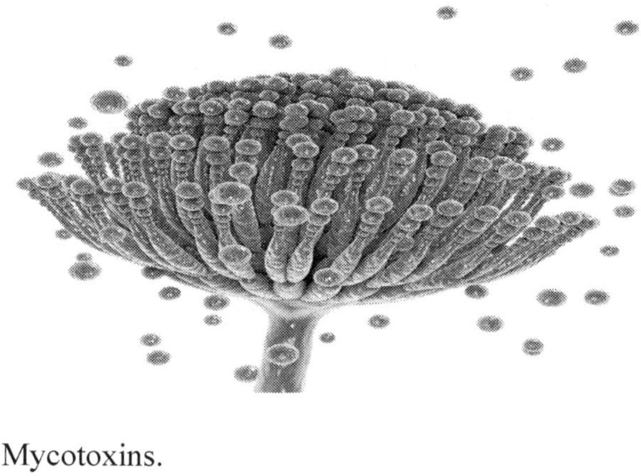

Figure 10.17: Mycotoxins.

http://www.envirologix.com/wp-content/uploads/2017/02/atoxin-red-sq-1.jpg

Mycotoxins: Foodstuff Protection Viewpoint

Mycotoxins are of main concern as of the provisions protection viewpoint comprise the ochratoxin A, aflatoxins, trichothecenes, zearalenone as well as fumonisins. Several mycotoxins like trichothecene stay toxic yet after ripening.

Mycotoxins may pose a diversity of unfavorable physical condition impacts in individuals. Further mycotoxins have the variety of additional fitness impacts comprising gastrointestinal disruptions, kidney disorder, reproductive contamination or disorders of the immune arrangement.

ACRYLAMIDES AND DIOXINS

DioxinS

Dioxins are the cluster of chemically-related compounds so as to be chiefly by-products of developed routes, however, are able to and consequence as of innate methods, for instance, volcanic outbreaks as well as wood blazes. Above 90% of the human being's contact is by foodstuff, chiefly animal protein as well as dairy foodstuffs, shellfish as well as fish. Limited disclosure of human beings to elevated concentrations of dioxins might consequence in skin scratches.

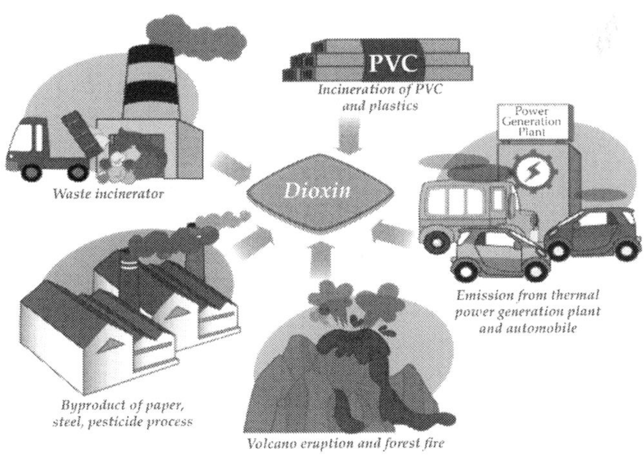

Figure 10.18: Dioxins.

http://tunza.eco-generation.org/editorPhoto/dioxin_28.gif

These are extremely poisonous as well as may pose developmental as well as reproductive disruptions, harm the immune system, obstruct with hormones Moreover also poses malignancy.

Dioxin: Reduction of Human Contact

Reduction otherwise prevention of human being introduction is best completed by means of source-directed procedures, that is a severe organization of developed procedures to diminish production of dioxins. This checking has led to the premature recognition of contagion as well as has repeatedly prohibited effect on a larger level.

Acrylamide: Production

Acrylamide is a substance that is present in the diversity of foodstuffs so as to have been processed as well as cooked at elevated temperatures, in addition to the intensity of acrylamide enhance by the duration of warmth. Though, in foodstuff, the means of production of acrylamide are improperly learned.

Figure 10.19: Acrylamide.

http://www.foodnavigator.com/var/plain_site/storage/images/publications/food-beverage-nutrition/foodnavigator.com/financial-industry/acrylamide-a-scandal-in-the-making/7743855-1-eng-GB/Acrylamide-A-scandal-in-the-making.jpg

During Cooking Process: Acrylamides Production

High concentrations of acrylamide produced while baking otherwise frying of cereal products or potatoes, uncovered by Swedish learnings in 2002. Learnings in laboratory animals recommended acrylamide has a carcinogenic effectiveness in rats so as to be comparable to that of further carcinogens within foodstuff, although the ingestion levels in favor of acrylamide are expected to be elevated.

Acrylamides Adverse Effect: Prevention

Currently, for acrylamide in provisions there are no regulatory utmost concentrations. To reduce risks following are the values might be functional whatever exists:

- Foodstuff must not be cooked enormously, that is for very long otherwise at too elevated a heat. Though, the entire foods, chiefly animal protein as well as meat foodstuffs, ought to be ripe carefully and completely to demolish foodborne microbes.
- People must consume a varied as well as balanced diet, which consists of abundance of vegetables as well as fruits, in addition to ought to modest their expenditure of fatty as well as fried provisions.

HAZARD AND HEALTH RISKS RELATED TO FOOD

Foodborne Illnesses

One of the most common illnesses because of hazard potential of chemical residual is following:

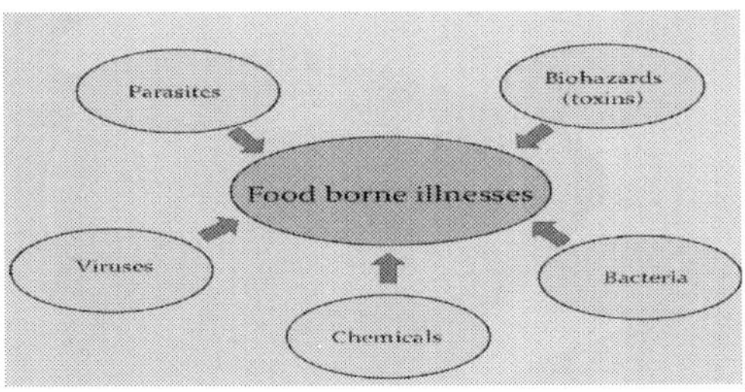

Figure 10.20: Sources of Foodborne Illnesses.

https://image.slidesharecdn.com/finalppt-140123214208-phpapp01/95/food-borne-animal-parasites-viruses-and-food-borne-biohazards-3-638.jpg?cb=1390513468

Foodstuff Poisoning

When definite pathogens otherwise disease-causing bacteria infect foodstuff, they are able to pose foodborne infirmity, frequently known as "food poisoning". Subtanences that are impure might not seem, smell or else taste any special as of sustenances so as to harmless to consume. Campylobacter, Salmonella, Escherichia coli (E. coli) as well as Listeria are general microbes posing foodborne disease. Several foodborne ailments are mild wherever the majority of the people obtain enhanced in some times, nevertheless infirmity may occasionally be additional harsh, even lethal.

Foodborne Infirmity: Clinical Symptoms

Following symptoms as well as signs are caused by most categories of foodborne illness:

- Vomiting
- Cramps and abdominal ache
- Nausea

Fever and/or the existence of mucus otherwise blood in the stool are symptoms of microbial contagion so as to might necessitate a medicinal discussion.

Foodborne Infirmity: Protection

Most foodborne diseases as well as illness can be prevented by suitable foodstuff treatment. WHO's five solutions to safe sustenance are as subsequent:

- Maintain Freshness:
    - Completely clean unprocessed vegetables as well as fruits by means of tap water.
    - Maintain hygienic kitchen, chopping board as well as hands every time.
- Isolate cooked food from raw food:
    - Don't merge ready-to-eat food as well as raw food.

    - Don't mix up raw fish, meat in addition to unprocessed vegetables.
- Complete cooking:
    - All poultry, meat as well as seafood, particularly shellfish, cook thoroughly.
    - Heat up the entire leftovers until they are blistering warm.
- Maintain foodstuff at protected warmth:
    - Within two hours of preparation, refrigerate cooked foodstuff.
    - At room temperature, never melt provisions. Unfreeze freezing provisions in the refrigerator, in the microwave otherwise in the frosty water.
- Apply safe raw materials as well as water.
    - For foodstuff preparation, apply protected drinking water.
    - Check labels as well as use-by dates during buying packed foodstuff.

REFERENCES

1. Abdel WA, Dawood RM, Abd E, Maha A, Helal S, *et al.*(2004) Detection of organochlorine pesticide residues in samples of cow milk collected from Sohag and Qena governorates. Ass UnivBull Environ Res 7: 105-116.
2. Abebew DS (2008) Detection and Determination of Oxytetracycline and Penicillin G Antibiotic Residue Levels In Bovine Bulk Milk from Debrezeit and Nazareth Dairy Farms. Proceedings of the 1st International Technology, Education, and Environment Conference.
3. Admasu E, Chandravanshi BS, Wondimu T, Akalu G (2005) Levels of some nutrients and non-essential metals in cow's whole milk delivered to Addis Ababa, Ethiopia. MSc Thesis, School of Graduate Studies, Addis Ababa University, Addis Ababa, Ethiopia.
4. AG (Australian Government) (2005) Acceptable daily intakes for agricultural and veterinary chemicals. Office of Chemical Safety, Department of Health and Ageing. Canberra, Australia 1–115.
5. Ahearn TU, McCullough ML, Flanders WD, Long Q, Sidelnikov E, Fedirko V, Daniel CR, Rutherford RE, Shaukat A, Bostick RM. 2011. A randomized clinical trial of the effects of supplemental

calcium and vitamin D_3 on markers of their metabolism in normal mucosa of colorectal adenoma patients. *Cancer Res***71**:413–23.

6. Ahn J, Albanes D, Peters U, Schatzkin A, Lim U, Freedman M, Chatterjee N, Andriole GL, Leitzmann MF, Hayes RB. 2007. Dairy products, calcium intake, and risk of prostate cancer in the prostate, lung, colorectal, and ovarian cancer screening trial. *Cancer Epidemiol Biomarkers Prev* **16**:2623–30.
7. Alamdar A, Syed JH, Malik RN, Katsoyiannis A, Liu J, Li J, Zhang G, Jones KC (2014) Organochlorine pesticides in surface soils from obsolete pesticide dumping ground in Hyderabad City, Pakistan: contamination levels and their potential for air–soil exchange. Sci Total Environ 470:733–741.
8. Alamgir Zaman Chowdhury M, Fakhruddin ANM, Nazrul Islam MD, Moniruzzaman M, Gan SH, Khorshed Alam MD (2013) Detection of the residues of nineteen pesticides in fresh vegetable samples using gas chromatography–mass spectrometry. Food Control 34:457–465.
9. Amin NU, Hussain A, Alamzeb S, Begum S (2013) Accumulation of heavy metals in edible parts of vegetables irrigated with waste water and their daily intake of adults and children, District Mardan, Pakistan. Food Chem 136(3):1515–1523.
10. Arain MB, Kazi TG, Baig JA, Jamali MK, Afridi HI, Shah AQ, Sarfraz RA (2009) Determination of arsenic levels in lake water, sediment, and foodstuff from selected area of Sindh, Pakistan: estimation of daily dietary intake. Food Chem Toxicol 47(1):242–248.
11. Aso Y, Akaza H, Kotake T, Tsukamoto T, Imai K, Naito S. 1995. Preventive effect of a Lactobacillus casei preparation on the recurrence of superficial bladder cancer in a double-blind trial. The BLP Study Group. *Eur Urol* **27**:104–9.
12. Alonso-Amelot, ME.; Castillo, U.; Smith, BL. & Lauren, DR. (1996). Bracken ptaquiloside in milk. Nature, Vol. 382, No.6592, pp 587, ISSN 0028-0836 Asociación Española de Farmacéuticos de la Industria. (March 2001).
13. Bonzongo JC, Donkor AK, Nartey VK, Lacerda LD (2004). Mercury pollution in Ghana: a case study of environmental impacts

of artisanal gold mining in sub-Saharan Africa. Lacerda LD, Santelli RE Facets of Environmental Geochemistry in Tropical and Sub-tropical Environments Springer Verlag Berlin 135 153.

14. Bal dit Sollier C, Drouet L, Pignaud G, Chevallier C, Caen J, Fiat AM, Izquierdo C, Jolles P. 1996. Effect of kappa-casein split peptides on platelet aggregation and on thrombus formation in the guinea-pig. *Thromb Res* **81**:427–37.
15. Beheshtipour H, Mortazavian AM, Haratian P, Darani KK. 2012. Effects of *Chlorella vulgaris* and *Spirulina platensis* addition on the viability of probiotic bacteria in yogurt and its biochemical properties. *Eur J Food Res Technol*, DOI: 10.1007/s00217-012-1798-4.
16. Biebinger R, Hurrell R, Ottaway P. 2008. Vitamin and mineral fortification of foods. In: Ottaway PB, editor. *Food fortification and supplementation: technological, safety and regulatory aspects.* Boca Raton, Fla.: CRC Press. p 27–40.
17. Barthelemy E, Spyropoulos D, Milana MR, *et al.* 2014. Safety evaluation of mechanical recycling processes used to produce polyethylene terephthalate (PET) intended for food contact applications. Food Addit Contam A. 31:490-7.
18. Brambilla, G.; Di Beza, S.; Pietraforte, D.; Minetti, M.; Campanella, L. & Loizzo, A. (2007) Oocyte quality and embryo production in cattle. Canadian 27 Journal of Animal Science, Vol. 78, (October 1998), pp. 513–516, ISSN 0008-3984.
19. Bower JR, Congeni BL, Cleary TG, Stone RT, Wanger A, Murray BE, et al. *Escherichia coli* O114: nonmotile as a pathogen in an outbreak of severe diarrhea associated with a day care center. J Infect Dis. 1989;160:243–247.
20. Christophersen, A. B., D. Levin, L. C. G. Hoegberg, H. R. Angelo & J. P. Kampmann: Activated charcoal alone or after gastric lavage: Early presenting intoxications with large ingestions, a volunteer study. *Brit. J. Clin. Pharmacol.* 2002, **53**, 312–317.
21. Campos LC, Whitam TS, Gome TAT, Andrad JRC, Trabulsi IR. *Escherichia coli serogroup* O111 includes several clones of diarrhagenic strains with different virulence properties. Infect Immun. 1994;62:3282–3288.

22. Carpten J, Nupponen N, Isaacs S, Sood R, Robbins C, Xu J, Faruque M, Moses T, Ewing C, Gillanders E. 2002. Germline mutations in the ribonuclease L gene in families showing linkage with HPC_1. *Nat Genet* **30**:181–4.

23. Chan JM, Pietinen P, Virtanen M, Malila N, Tangrea J, Albanes D, Virtamo J. 2000. Diet and prostate cancer risk in a cohort of smokers, with a specific focus on calcium and phosphorus (Finland). *Cancer Causes Control* **11**:859–67.

24. Cho E, Smith-Warner SA, Spiegelman D, Beeson WL, van den Brandt PA, Colditz GA, Folsom AR, Fraser GE, Freudenheim JL, Giovannucci E. 2004a. Dairy foods, calcium, and colorectal cancer: a pooled analysis of 10 cohort studies. *J Natl Cancer Inst* **96**:1015–22.

25. Cruz AG, Cadena RS, Walter EHM, Mortazavian AM, Granato D, Faria JAF, Bolini H. 2010. Sensory analysis: relevance for prebiotic, probiotic, and synbiotic product development. *Compr Rev Food Sci Food Saf* **9**:358–73.

26. Chamorro-Garcia R, Kirchner S, Li X, *et al.* 2012. Bisphenol A diglycidyl ether induces adipogenic differentiation of multipotent stromal stem cells through a peroxisome proliferator-activated receptor gamma-independent mechanism. Environ Health Persp. 120:984-9.

27. Cleveland et al., 2003 T.E. Cleveland, P.F. Dowd, A.E. Desjardins, D. Bhatnagar, P.J. Cotty Pest Manage. Sci., 59 (2003), pp. 629-642 Cook *et al.*, 1986 W.O. Cook, J.L. Richard, G.D. Osweiller, D.W. Trampel Am. J. Vet. Res., 47 (1986), pp. 1817-1825.

28. Cullen and Newberne, 1994 J.M. Cullen, P.N. Newberne D.L. Eaton, J.J. Groopman (Eds.), The Toxicity of Aflatoxins. Human Health, Veterinary and Agricultural Significance, Academic Press, San Diego (1994), pp. 3-26.

29. Commission Regulation (EC) No 629/2008 of 2 July 2008 amending Regulation (EC) No 1881/2006 setting maximum levels for certain contaminants in foodstuffs S.I. No. 389 of 1993. Health (Tin in Food) Regulations, 1993 S.I. No. 44 of 1972. Health (Arsenic in Food) Regulation, 1972 S.I. No. 72 of 1992. Health (Arsenic in Food) (Amendment) Regulations 1992 S.I. No. 400 of 2001.

30. Commission Directive 2005/31/EC of 29 April 2005 amending Council Directive 84/500/EEC as regards a declaration of compliance and performance criteria of the analytical method for ceramic articles intended to come into contact with foodstuffs, (L110), p36, 30/04/2005.
31. Chemical residues in meat and meat products: a review (pdf download available). Available from: https://www.researchgate.net/publication/289614526_CHEMICAL_RESIDUES_IN_MEAT_AND_MEAT_PRODUCTS_A_REVIEW [accessed Aug 7, 2017].
32. Council Directive 84/500/EEC of 15 October 1984 on the approximation of the laws of the Member States relating to ceramic articles intended to come into contact with foodstuffs; Official Journal of the European Communities, (L277), p12, 20/10/1984 Council Directive 98/83/EC of 3 November 1998 on the quality of water intended for human consumption.
33. Carryover of pyrrolizidine alkaloids from feed to milk in dairy cows. Food Additives and Contaminants. Part A Chemistry, analysis, control, exposure and risk assessment, Vol.28, No.3, (March, 2011), pp 359-372, ISSN 1944-0049 International Agency for Research on Cancer, IARC. (1993).
34. Cudjoe AR, Kyofa-Boamah M, Braun M (2002) Handbook of crop protection recommended in Ghana. An IPM Approach. Vol.4: Selected fruit crops (mango, papaya and pineapple). Ministry of Food and Agriculture/Plant Protection and Regulatory Services Directorate/GTZ. Accra-Ghana. pp 60–63.
35. Dawd AG, Gezmu TB, Haki GD (2012) Essential and Toxic Metals in cow's whole milk from selected sub-cities in Addis Ababa, Addis Ababa University, Center for Food Science and Nutrition, Ethiopia. Inter J of Food Science 1: 12-19.
36. Determination of Clenbuterol in Pig Liver Following Prolonged Administration of a Growth-Promoting Dose. Food Technology and Biotechnology, Vol. 40, No. 4, (November 2002), pp. 343–346, ISSN 1330-9862.
37. Danaher, M.; De Ruyckb, H.; Crooks, S.; Dowling, G. & O'Keeffe, M. (2007)Experimental selection of mebendazole resistant gill

monogeneans from European eel Anguilla anguilla. Journal of Fish Diseases, Vol. 15 No. 5, (September 1992), pp. 393-400, ISSN 1365-2761.

38. Dorgan JF, Judd JT, Longcope C, Brown C, Schatzkin A, Clevidence BA, Campbell WS, Nair PP, Franz C, Kahle L. 1996. Effects of dietary fat and fiber on plasma and urine androgens and estrogens in men: a controlled feeding study. *Am J Clin Nutr* **64**:850–5.
39. Eroksuz, Y.; Ceribasi, A.; Cevik, A.; Eroksuz, H.; Tosun, F. & Tamer, U. (2008)Paralytic shellfish poisoning: Sea food safety and human health perspectives. Toxicon, Vol. 56, No.2, (August, 2010), pp 108–122, ISSN 1879-3150.
40. European Food Safety Authority (EFSA). 2010. Scientific Opinion on the re-evaluation of Brilliant Blue FCF (E 133) as a food additive. *EFSA Panel on Food Additives and Nutrient Sources added to Food. EFSA J 8* **(11)**: 1853; 1–36.
41. Elwood PC, Givens DI, Beswick AD, Fehily AM, Pickering JE, Gallacher J. 2008. The survival advantage of milk and dairy consumption: an overview of evidence from cohort studies of vascular diseases, diabetes and cancer. *J Am Coll Nutr* **27**:723S–34S.
42. Europian Commission (1997) Regulation Laying Down A Community Procedure for the Establishment of Maximum Residue Limits of Veterinary Medical Products in Foodstuffs of Animal Origin. Offic J 224: 1-8.
43. E. Lombi and M. H. Gerzabek, "Determination of mobile heavy metal fraction in soil: results of a pot experiment with sewage sludge," Communications in Soil Science and Plant Analysis, vol. 29, no. 17-18, pp. 2545–2556, 1998.
44. Eqani SAMAS, Malik RN, Katsoyiannis A, Zhang G, Chakraborty P, Mohammad A, Jones KC (2012a) Distribution and risk assessment of organochlorine contaminants in surface water from River Chenab, Pakistan. J Environ Monit 14:1645–1654.
45. Eqani SAMAS, Malik RN, Alamdar A, Faheem H (2012b) Status of organochlorine contaminants in the different environmental compartments of Pakistan: a review on occurrence and levels. Bull Environ Contam Toxicol 88:303–10.

46. Eqani SAMAS, Malik RN, Cincinelli A, Zhang G, Mohammad A, Qadir A, Rashid A, Bokhari H, Jones KC, Katsoyiannis A (2013) Uptake of organochlorine pesticides (OCPs) and polychlorinated biphenyls (PCBs) by river water fish: the case of River Chenab. Sci Total Environ 450:83–91.

47. European Commission Regulation 333/2007 laying down the sampling methods and the methods of analysis for the official control of the lead, cadmium, mercury, inorganic tin, 3-MCPD and benzo(a)pyrene in foodstuffs.

48. European Communities (Sampling Methods and Methods of analysis for the official control of the levels of certain contaminants in foodstuffs) Regulations 2006 Rose, M., Knaggs, M., Owen, L., and Baxter, M., (2001).

49. Food and Agricultural Organization (FAO) (2002) FAO/WHO global forum of food safety regulators. Marrakech, Morocco Marrakech, Morocco, 28–30 January 2002 Country Report proposed by Iran. http://www. fao.org/DOCREP/MEETING/004/AB428E.HTM Agenda Item 4.2 a, GF/CRD Iran-1. Accessed 7 April 2016.

50. Fletouris DJ, Botsoglou NA, Psomas IE, Mantis AI (1997) Determination of the Marker Residue of Albendazole in Cheese by Ion-pair liquid Chromatography and Fluorescence Detection. J Dairy Sci 80: 2695-2700.

51. Farlow DW, Xu X, Veenstra TD. 2009. Quantitative measurement of endogenous estrogen metabolites, risk-factors for development of breast cancer, in commercial milk products by LC-MS/MS. *Chromatogr B Anal Technol Biomed Life Sci* **877**(13):1327–34.

52. Fedirko V, Bostick RM, Flanders WD, Long Q, Sidelnikov E, Shaukat A, Daniel CR, Rutherford RE, Woodard JJ. 2009. Effects of vitamin d and calcium on proliferation and differentiation in normal colon mucosa: a randomized clinical trial. *Cancer Epidemiol Biomarkers Prev* **18**:2933–41.

53. Fagundes Neto U, Ferreira V, Patricio FRS, Mostaço VL, Trabulsi LR. Protracted diarrhea: the importance of the enteropathogenic *E. coli* (EPEC) strains and *Salmonella* in its genesis. J Pediatr Gastroenterol Nutr. 1989;8:207–211. doi: 10.1097/00005176-198902000-00014.

54. Fischer WJ, Schilter B, Tritscher AM, Stadler RH. 2011. Contaminants of milk and dairy products. In: *Contamination resulting from farm and dairy practices: encyclopedia of dairy sciences*. 2nd ed. San Diego: Academic Press. p 887–97.

55. Golan E, Krissoff B, Kuchler F, Calvin L, Nelson K, Price G. 2004. Traceability in the U.S. food supply: economic theory and industry studies. Agricultural economic report nr 830. Washington, DC: Economic Research Service, U.S. Dept. of Agriculture. 48 p.

56. Gennaro, M. C.; Abrigo, C. (1992). "Caffeine and theobromine in coffee, tea and cola-beverages or any other fizzy drinks". Fresenius' Journal of Analytical Chemistry. **343** *(6): 523–525. ISSN 0937-0633. doi:10.1007/BF00322162.*

57. Gebremichael S, Birhanu T, Tessema DA (2013) Analysis of Organochlorine Pesticide Residues in human and cow's milk in the towns of Asendabo, Serbo and Jimma in South-Western Ethiopia. ChemPubl 5:1652-1657.

58. Guidelines for the application of the regulation (EC) 2023/2006 to the supply chain of materials and articles intended to come into contact with food (Italian guideline, in English, http://www.iss.it/binary/publ/cont/ 11_37_web.pdf).

59. G. Sposito and A. L. Page, "Cycling of metal ions in the soil environment," in Metal Ions in Biological Systems, H. Sigel, Ed., vol. 18 of Circulation of Metals in the Environment, pp. 287–332, Marcel Dekker, Inc., New York, NY, USA, 1984.

60. H. Gans, Joseph. "Effects of short-term and long-term theobromine administration to male dogs". Toxicology and Applied Pharmacology. **53**: 481–496. doi:10.1016/0041-008X(80)90360-9. Retrieved 1 March 2012.

61. Hoegberg, L. C. G., H. R. Angelo, A. B. Christophersen & H. R. Christensen: Effect of ethanol and pH on the adsorption of paracetamol to high surface activated charcoal, *in vitro* studies. *J. Toxicol. Clin. Toxicol.* 2002, **40**, 59–67.

62. Hunt RG, Sellers VR, Frankalin WE, Nelson JM, Rathje WL, Hughes WW, Wilson DC. 1990. Estimates of the volume of MSW and selected components in trash cans and land fills. Tucson , Ariz.:

Report prepared by The Garbage Project and Franklins Assn. Ltd. for the Council for Solid Waste Solutions.

63. Hjorth K, Johansen K, Holen B, Andersson A, Christensen HB, Siivinen K, *et al.* (2011) Pesticide residues in fruits and vegetables from South America. A Nordic project. Food Control 22:1701–1706.

64. Iamiceli, A. L., I. Fochi, G. Brambilla and A. Di Domenico. (2009). Determination of persistant organic pollutants in meat. In Handbook of Processed Meats and Poultry Analysis, edited by L. M. L. Nollet and F. Toldra. Boca Raton, Fla.: CRC Press.

65. Jahed KGR (2005) A Model for Application of Hazard Analysis Critical Control Point (HACCP) system in production of Iranian white cheese stored in salt brine. Proceedings of the 8th National Congress on Environmental Health, Tehran, Iran, pp: 333.

66. J. Bjuhr, Trace Metals in Soils Irrigated with Waste Water in a Periurban Area Downstream Hanoi City, Vietnam, Seminar Paper, Institutionen för markvetenskap, Sveriges lantbruksuniversitet (SLU), Uppsala, Sweden, 2007.

67. J. Buekers, Fixation of cadmium, copper, nickel and zinc in soil: kinetics, mechanisms and its effect on metal bioavailability, Ph.D. thesis, Katholieke Universiteit Lueven, 2007, Dissertationes De Agricultura, Doctoraatsprooefschrift nr.

68. K. Weggler, M. J. McLaughlin, and R. D. Graham, "Effect of Chloride in Soil Solution on the Plant Availability of Biosolid-Borne Cadmium," Journal of Environmental Quality, vol. 33, no. 2, pp. 496–504, 2004.

69. Kyofa-Boamah M, Blay E (2000) A study on pineapple production and protection procedure in Ghana. Ministry of Food and Agriculture, Plant Protection and Regulatory Services Directorate. Accra, Ghana. pp 23.

70. Lozano, MC., & Arias, DC. (2008). Biochemistry of hemlock (Conium maculatum L.) alkaloids and their acute and chronic toxicity in livestock. A review. Toxicon, Vol.37, No.6, (June, 1999), pp 841-865, ISSN 1879-3150.

71. Lyon, France Jiménez, V.; Companyó, R. & Guiteras, J. (2011).

Some Naturally Occurring Substances: Food Items and Constituents, Heterocyclic Aromatic Amines and Mycotoxins. IARC Monographs on the evaluation of carcinogenic risk to humans 56. IARC, pp. (1-599) ISBN 92 832 12568.

72. Lehane, L. & Lewis, RJ. (2000). Effects of zearalenone and α-Zearalenol in comparison with Raloxifene on T47D cells. Toxicology mechanisms and methods, Vol. 19, No. 3, (March, 2009), pp 245-250, ISSN 1537-6516.
73. Lewis *et al.*, 2005 L. Lewis, M. Onsongo, H. Njapau, H. Schurz Rogers, G. Luber, S. Kieszak, J.Nyamongo, L. Backer, A.M. Dahiye, A. Misore, K. Decoct, C. Rubin Environ. Health Perspect., 113 (2005), pp. 1763-1767.
74. Lipton, S. H., C. E. Bodwell and A. H. Coleman Jr. Amino acid analyzer studies of the products of peroxide oxidation of cystine, lanthionine and homocystine. Journal of Agriculture and Food Chemistry, 1977; 25: 624 – 628.
75. Masud SZ, Hassan N (1995) Pesticides residues in fruits in food stuffs in Pakistan Organochlorine, organophosphates and pyrethroids insecticides in fruits and vegetables. Pak J Sci Ind Res 35:499–504.
76. Medina, M.; González D. & Ramírez A. (2008). Multiresidue analysis of antibiotics in food of animal origin using liquid chromatography-mass spectrometry. Methods in Molecular Biology, Vol. 747, (May 2011), pp. 267-307, ISSN 1940-6029.
77. M. J. McLaughlin, R. E. Hamon, R. G. McLaren, T. W. Speir, and S. L. Rogers, "Review: a bioavailability-based rationale for controlling metal and metalloid contamination of agricultural land in Australia and New Zealand," Australian Journal of Soil Research, vol. 38, no. 6, pp. 1037–1086, 2000.
78. Moreno, L.; Alvarez, L.; Ceballos, L.; Sánchez Bruni S. & Lanusse C. (2008). Effects of an Estradiol Implant on Locoweed Consumption, Toxicity, and Recovery in Growing Beef Steers, Professional Animal Scientist, Vol. 17, No. 2, (June 2001), pp. 109-11, ISSN 1080-7446.
79. Marsh KS. 1993. Packaging as a supplier of health, filler of landfills, and excuse for trade barriers [abstract]. In: International symposium on packaging, economic development and the environment, 1992

Nov 9, Chicago Ill. Herndon, Va .: Inst. of Packaging Professionals . p 23–36.

80. McDonough W, Braungart M. 2002. Cradle to cradle: remaking the way we make things. New York : North Point Press. 212 p. McKown C. 2000. Containers. In: Coatings on glass—technology roadmap workshop. 2000 Jan 18–19. Livermore , Calif : Sandia National Laboratories report . p 8–10.
81. Mallat E, Barzen C, Abuknesha R, Gauglitz G, Barcelo D (2001) Fast determination of paraquat residues in water by an optical immunosensor and validation using capillary electrophoresis-ultraviolet detection. J Anal ChimActa 427: 165-171.
82. Mokwunye U (1981). The effects of the application of 3 phosphate fertilizers: Ground rock phosphate (GRP), Ground Rock Phosphate Sulphur (GRP+S) and single superphosphate (SSP) on the amounts and ratios of exchangeable cations in Nigerian savanna soils. Samaru J. Agric. Res. 1:23-34.
83. Metals and alloys: Council of Europe (2013): Metals and alloys used in food contact materials and articles. A practical guide for manufactures and regulators. Committee of experts on packaing materials for food and pharmaceutical products. (P-SC-EMB) 1–215.
84. Moats, W. A. (1994). Chemical residues in muscle foods. In Muscle Foods: Meat, Poultry and Seafood Technology, edited by D. M. Kinsman, A. W. Kotula and B. C. Breidenstein. New York: Chapman and Hall.
85. Nakata H, Kawazoe M, Arizono K, Abe S, Kitano T, Shimada H, Ding X (2002) Organochlorine pesticides and polychlorinated biphenyl residues in foodstuffs and human tissues from China: status of contamination, historical trend, and human dietary exposure. Arch Environ Contam Toxicol 43(4):473–480.
86. N. S. Bolan, B.G. Ko, C.W.N. Anderson, and I. Vogeler, "Solute interactions in soils in relation to bioavailability and remediation of the environment," in Proceedings of the 5th International Symposium of Interactions of Soil Minerals with Organic Components and Microorganisms, Pucón, Chile, November 2008.

87. Osei Tutu A, Yeboah PO, Golow AA, Denutsui D, Blankson-Arthur S (2011) Organochlorine pesticides residues in the breast milk of some Primiparae mothers in La Community, Accra, Ghana, res. J Environ Earth Sci 3(2):153–159.

88. Opinion of the Scientific Panel on Contaminants in the Food Chain on a request from the European Commission on glucosinolates as undesirable substances in animal feed, The EFSA Journal 590, pp 1-76, ISSN 1831-4732 European Agency for the Evaluation of Medicinal Products, EMEA. (1997).

89. P. S. DeVolder, S. L. Brown, D. Hesterberg, and K. Pandya, "Metal bioavailability and speciation in a wetland tailings repository amended with biosolids compost, wood ash, and sulfate," Journal of Environmental Quality, vol. 32, no. 3, pp. 851–864, 2003.

90. Pyrrolizidine alkaloids in food: a spectrum of potential health consequences. Food Addittives and Contaminants. Part A chemistry, analysis, control, exposure and risk assessment, Vol.28, No.3, (March 2011), pp 308–324, ISSN: 1944-0049.

91. Rathje WL, Reilly MD, Hughes WW. 1985. Household garbage and the role of packaging—the United States/Mexico City household refuse comparison. Tucson , Ariz. : Solid Waste Council of the Paper Industry Publishing. p 116.

92. R. L. Chaney and D. P. Oliver, "Sources, potential adverse effects and remediation of agricultural soil contaminants," in Contaminants and the Soil Environments in the Australia-Pacific Region, R. Naidu, Ed., pp. 323–359, Kluwer Academic Publishers, Dordrecht, The Netherlands, 1996.

93. R. G. McLaren, L. M. Clucas, and M. D. Taylor, "Leaching of macronutrients and metals from undisturbed soils treated with metal-spiked sewage sludge. 3. Distribution of residual metals," Australian Journal of Soil Research, vol. 43, no. 2, pp. 159–170, 2005.

94. Report from Task 3.2.11: Assessment of the dietary exposure to arsenic, cadmium, lead and mercury of the population of the EU Member States. European Commission, Directorate- General Health and Consumer Protection. SCOOP report EFSA (2005).

95. Regulation EU 284/2011 laying down specific conditions and detailed procedures for the import of polyamide and melamine plastic kitchenware originating in or consigned from the People's Republic of China and Hong Kong Special Administrative Region, China.
96. R. G. McLaren, L. M. Clucas, M. D. Taylor, and T. Hendry, "Leaching of macronutrients and metals from undisturbed soils treated with metal-spiked sewage sludge. 2. Leaching of metals," Australian Journal of Soil Research, vol. 42, no. 4, pp. 459–471, 2004.
97. Staffeld R, and Kulke E. 2011. Informal employment and health conditions in Dhaka's plastic recycling and processing industry. In: Health in megacities and urban areas. A. Krämer, M.H. Khan and F. Kraas, eds. Springer Verlag, Berlin Heidelberg.
98. Shephard, 2006 G.S. Shephard D. Barug, D. Bhatnagar, H.P. Van Egmond, J.W. Van der Kamp, W.A. Van Ossenbruggen, A. Visconti (Eds.), The Mycotoxin Fact Book, Wageningen Academic,Wageningen (2006), pp. 21-36.
99. Talanta, Vol. 85 No. 1, (July 2011), Validation of a method for the analysis of nine quinolones in eggs by pressurized liquid extraction and liquid chromatography with fluorescence detection. pp. 596-606. ISSN 0039-9140 Khosrokhavar, R.; Rahimifard, N.; Shoeibi S.; Pirali, M. & Hosseini, M. (2009).
100. T. A. Kirpichtchikova, A. Manceau, L. Spadini, F. Panfili, M. A. Marcus, and T. Jacquet, "Speciation and solubility of heavy metals in contaminated soil using X-ray microfluorescence, EXAFS spectroscopy, chemical extraction, and thermodynamic modeling," Geochimica et Cosmochimica Acta, vol. 70, no. 9, pp. 2163–2190, 2006.
101. Tardieu, D.; Bailly, JD. Skiba, F.; Grosjean, F. & Guerre, P. (2008). Defining the neurotoxin derived illness chronic ciguatera using markers of chronic systemic inflammatory disturbances: a case/control study. Neurotoxicology and Teratology, Vol.32, No.6, (December, 2010), pp 633-639, ISSN 1872-9738.
102. Toxicity of Heliotropium dolosum, Heliotropium circinatum, and Senecio vernalis in Parental Quail and Their Progeny, with Residue

Evaluation of Eggs. Turkish Journal of Veterinary and Animal Sciences, Vol.32, No. 6, pp 475-482, ISSN 1303-6181 European Food Safety Authority, EFSA. (2008).

103. World Health Organization International Agency for Research on Cancer (WHO-IARC), 1993a World Health Organization International Agency for Research on Cancer (WHO-IARC), 1993a. IARC Monographs on the Evaluation of Carcinogenic Risks to Humans, vol. 56, pp. 445–462.

104. Waes, S.A. Missmer, J. Cabrera, O. Torres, W.C.A. Gelderblom, J.Allegood, C. Martinez, J. Maddox, J. Miller, L. Starr, M.C. Sullards, A. Roman, K.A.Voss, E. Wang, A.H. Merril J. Nutr., 134 (2004), pp. 711-722.

INDEX

G

H

I

L

M

N

O

P

R

S

T

U

V